Transition Metal-Catalyzed Pyridine Synthesis

Transition Metal-Catalyzed Pyridine Synthesis

Transition Metal-Catalyzed Heterocycle Synthesis Series

Xiao-Feng Wu
Department of Chemistry, Zhejiang Sci-Tech Univeristy, China
and
Leibniz-Institut für Katalyse e.V., Universität Rostock, Germany

ELSEVIER

AMSTERDAM • BOSTON • HEIDELBERG • LONDON
NEW YORK • OXFORD • PARIS • SAN DIEGO
SAN FRANCISCO • SINGAPORE • SYDNEY • TOKYO

Elsevier
Radarweg 29, PO Box 211, 1000 AE Amsterdam, Netherlands
The Boulevard, Langford Lane, Kidlington, Oxford OX5 1GB, UK
50 Hampshire Street, 5th Floor, Cambridge, MA 02139, USA

ISBN: 978-0-12-809379-5

British Library Cataloguing-in-Publication Data
A catalogue record for this book is available from the British Library

Library of Congress Cataloging-in-Publication Data
A catalog record for this book is available from the Library of Congress

For Information on all Elsevier Publishing
visit our website at http://store.elsevier.com/

This book has been manufactured using Print On Demand technology.

Dedicated to my wife, children, and my parents

Thanks for their understanding, support, encouragement, and tolerance!

CONTENTS

CONTENTS

CHAPTER 1

Introduction

Pyridine is an important class of nitrogen-containing heterocycles found in various natural products, pharmaceuticals, and materials (Scheme 1.1). Based on its importance, numerous synthetic procedures have been developed for their preparation [1]. In this book volume, the main achievements on transition metal-catalyzed pyridines synthesis are discussed. Based on the reaction types, the whole volume is catalogued by intramolecular cyclization and intermolecular cyclization reactions.

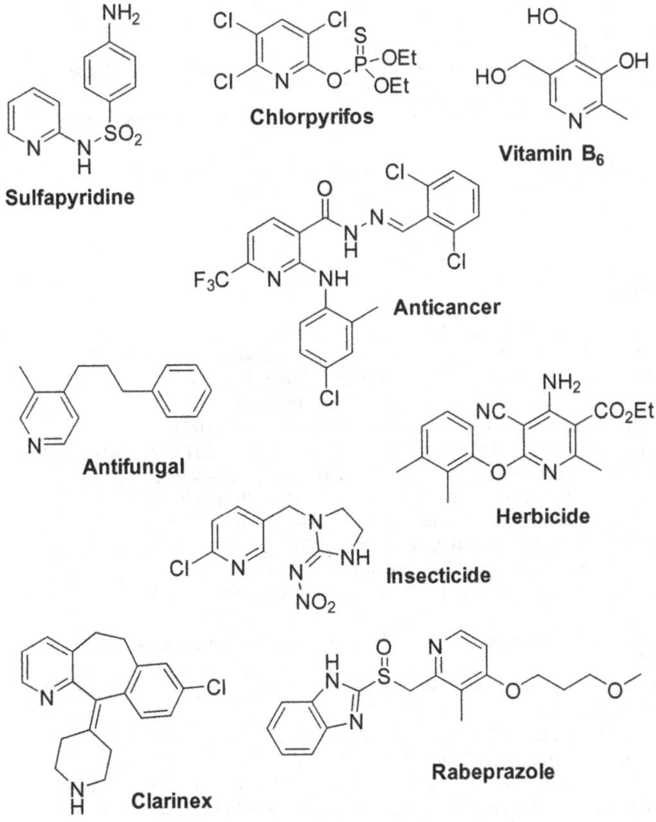

Scheme 1.1 Selected examples of bio-active pyridine derivatives.

Transition Metal-Catalyzed Pyridine Synthesis. DOI: http://dx.doi.org/10.1016/B978-0-12-809379-5.00001-9

Rosiglitazone   Pioglitazone

Lansoprazole   Pantoprazole

Etoricoxib

Anabasine   Imatinib Mesylate

Scheme 1.1 (Continued)

REFERENCE

[1] Reviews on pyridines synthesis see:

a. Gromov, S. P.; Fomina, M. V. *Russ. Chem. Rev.* **2008**, *77*, 1055–1077.
b. Chopade, P. R.; Louie, J. *Adv. Synth. Catal.* **2006**, *348*, 2307–2327.
c. Bönnemann, H. *Angew. Chem. Int. Ed. Engl.* **1985**, *24*, 248–262.
d. Heller, B.; Hapke, M. *Chem. Soc. Rev.* **2007**, *36*, 1085–1094.
e. Baumann, M.; Baxendale, I. R. *Beilstein J. Org. Chem.* **2013**, *9*, 2265–2319.
f. Weding, N.; Hapke, M. *Chem. Soc. Rev.* **2011**, *40*, 4525–4538.
g. Domínguez, G.; Pérez-Castells, J. *Chem. Soc. Rev.* **2011**, *40*, 3430–3444.
h. Neely, J. M.; Rovis, T. *Org. Chem. Front.* **2014**, *1*, 1010–1015.
i. Hill, M. D. *Chem. Eur. J.* **2010**, *16*, 12052–12062.
j. Varela, J. A.; Saá, C. *Chem. Rev.* **2003**, *103*, 3787–3801.
k. Bull, J. A.; Mousseau, J. J.; Pelletier, G.; Charette, A. B. *Chem. Rev.* **2012**, *112*, 2642–2713.
l. Gulevich, A. V.; Dudnik, A. S.; Chernyak, N.; Gevorgyan, V. *Chem. Rev.* **2013**, *113*, 3084–3213.
m. Allais, C.; Grassot, J.-M.; Rodriguez, J.; Constantieux, T. *Chem. Rev.* **2014**, *114*, 10829–10868.
n. Amatore, M.; Aubert, C. *Eur. J. Org. Chem.* **2015**, 265–286.
o. Henry, G. D. *Tetrahedron* **2004**, *60*, 6043–6061.
p. Kotha, S.; Brahmachary, E.; Lahiri, K. *Eur. J. Org. Chem.* **2005**, 4741–4767.
q. Pla-Quintana, A.; Roglans, A. *Molecules* **2010**, *15*, 9230–9251.
r. Varela, J.; Saá, C. *Synlett* **2008**, 2571–2578.

CHAPTER 2

Synthesized by Intramolecular Cyclizations

Intramolecular cyclization of substrates to the corresponding pyridine derivatives is the most straightforward pathway. By intramolecular version reaction, the positions of substituents are fixed and easier to control, such as N-propargylic β-enaminones in pyridines and pyrroles synthesis. In 2008, Cacchi et al. reported a copper-catalyzed synthesis of polysubstituted pyridines from N-propargylic β-enaminones [1]. By using DMSO as the solvent, pyrroles can be produced in good to high yields in the presence of only Cs_2CO_3 at room temperature, while pyridines can be formed under the assistant of CuBr at $60-80°C$ (Scheme 2.1). Here, the N-propargylic β-enaminones can be prepared via the following sequences: (1) cross-coupling of terminal alkynes with acyl chlorides; (2) followed by the conjugate addition of propargylamine with the resultant α,β-enones; (3) further Sonogashira cross-coupling of the propargyl derivative with aryl halides.

Scheme 2.1

A Carousel Tube Reactor (Radley Discovery), equipped with a magnetic stirrer, was charged with substrate (0.23 mmol) and recrystallized CuBr (13 mg, 0.09 mmol) in anhydrous DMSO (3 mL) was stirred for 1.5 h under argon at 60°C. Ethyl acetate was added to the reaction mixture and the resulting mixture was washed first with a 0.1 N HCl solution, then with a saturated NaCl solution, and finally dried over Na_2SO_4. The solvent was removed under vacuum and the residue was purified by chromatography (SiO_2, 30 g, n-Hexane/AcOEt 70/30 v/v) to give the corresponding pyridine derivative.

Transition Metal-Catalyzed Pyridine Synthesis. DOI: http://dx.doi.org/10.1016/B978-0-12-809379-5.00002-0

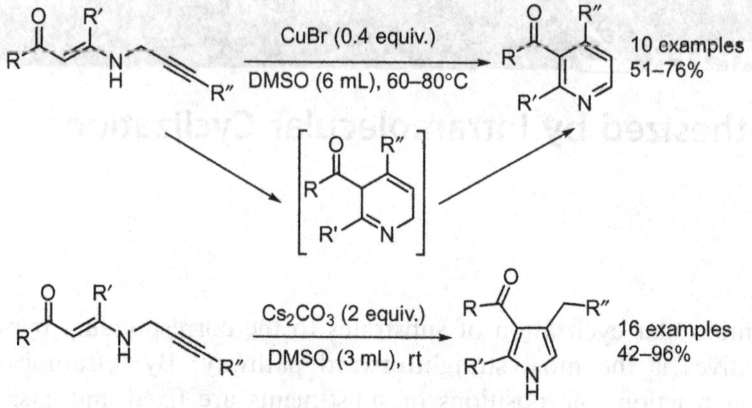

Scheme 2.1 *CuBr-catalyzed pyridines synthesis from N-propargylic β-enaminones.*

A procedure by using enamino ester and alkynone as the substrates was developed as well [2]. 2,3,4,6-Tetrasubstituted pyridines were prepared in a single step. Various acids, such as acetic acid, Amberlyst 15 ion exchange resin, zinc(II) bromide or ytterbium(III) triflate, can be applied as promoter for the cyclization step of the Michael addition adduct. 4-(3-Oxoalkyl)isoxazoles found could be applied as starting material for pyridine synthesis as well [3].

In 2013, Kim and coworkers reported a palladium-catalyzed domino cyclization of *N*-(2-bromoallyl)-*N*-cinnamyltosylamides for the construction of pyridines [4]. The reaction proceeds via a domino 5-exo/3-exo carbopalladation, ring expansion by palladium rearrangement, and an aromatization. Various 4-arylnicotinate derivatives were produced in good yields (Scheme 2.2).

Scheme 2.2

A mixture of substrate (0.5 mmol), Pd(OAc)$_2$ (6 mg, 5 mol%), PPh$_3$ (13 mg, 10 mol%), and Cs$_2$CO$_3$ (408 mg, 2.5 equiv.) in DMF (1.5 mL) was heated to 120°C for 3 h. The reaction mixture was poured into dilute HCl solution and extracted with diethyl ether. The organic solvent was removed by evaporation, and the residue was purified by flash column chromatography (hexanes/ether, 3:1) to afford the pure product.

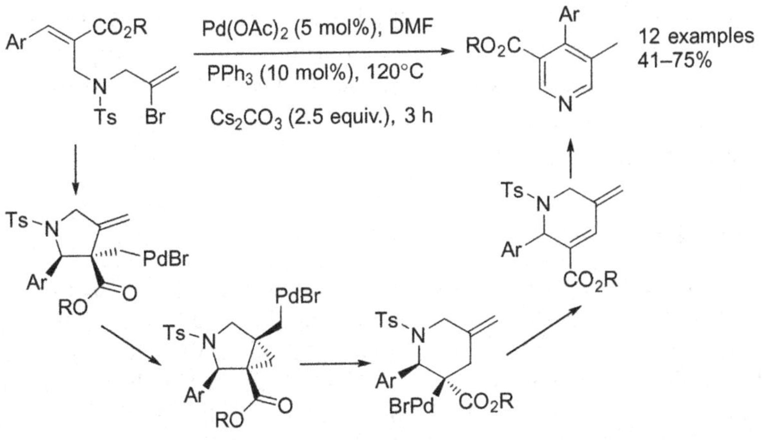

Scheme 2.2 Pd-catalyzed pyridines synthesis from N-(2-bromoallyl)-N-cinnamyltosylamides.

In 2014, Gagosz and coworkers reported their achievement on gold-catalyzed transformation of $2H$-azirines to the corresponding functionalized pyridines [5]. Good yields of polysubstituted functionalized pyridines were produced from easily accessible 2-propargyl $2H$-azirine derivatives (Scheme 2.3). High functional-group tolerance and wide substrate scope were exhibited. Later on, a DFT study was performed [6]. They show that the reaction was promoted by the gold(I) catalyst through 1,5-endo-dig cyclization by nucleophilic attack, direct ring expansion (C-N bond cleavage) followed by 1,2-proton transfer.

Scheme 2.3

To a solution of 0.10–0.20 mmol (1.0 equiv.) of the azirine in 1,2-dichloroethane (0.1 M) was added tBuXPhosAuNTf$_2$ (2 mol%). The solution was heated to reflux until TLC analysis indicated completion of the reaction (ca. 1–12 h). The solvent was removed under reduced pressure and the crude product was purified by column chromatography. The silica gel used for column chromatography was deactivated with diethylamine prior to use.

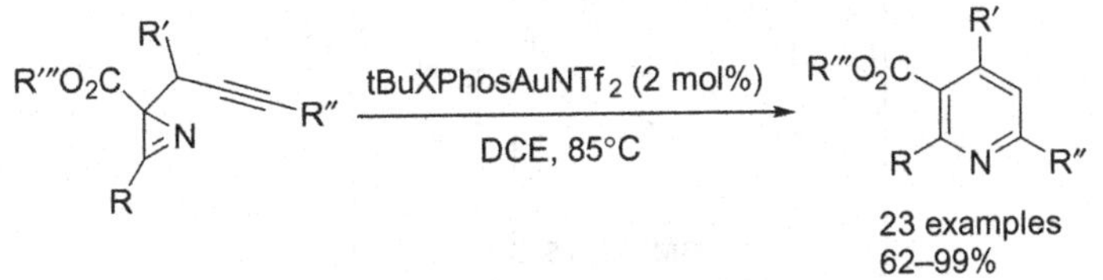

Scheme 2.3 Au-catalyzed pyridines synthesis from azirines.

Eastgate and coworkers developed a Ni-catalyzed C-H functionalization procedure for the synthesis of pyridine core and applied in the synthesis of the potent JAK2 inhibitor BMS-911543 [7]. With nickel oxide supported on silica (PRICAT) as the catalyst, 74% of the key intermediate can be achieved (Scheme 2.4).

Scheme 2.4

To a pressure vessel charged with hydroxyamidine (50.0 g, 0.164 mol, 1.0 equiv.) in AcOH (0.650 L, 13.0 mL/g) and PhMe (0.500 L, 10.0 mL/g) was added Piv$_2$O (30.5 g, 0.164 mol, 1.0 equiv.) followed by PhMe (0.150 L, 3.0 mL/g). The pressure vessel was degassed with N$_2$, and the reaction mixture was stirred at 20−30°C for 0.5 h. PRICAT (9.00 g, 18−20 wt%) as a slurry in AcOH (0.100 L, 2.0 mL/g) and PhMe (0.1 L, 2.0 mL/g) was added and the pressure vessel degassed with nitrogen and then pressurized with hydrogen (3 psig). The reaction mixture was stirred at 20−30°C for 20 h. Upon completion, the hydrogen was purged from the pressure vessel with nitrogen. The reaction mixture was filtered through a plug of Celite, and the cake was washed with PhMe (2 × 0.150 L, 2 × 3.0 mL/g). The reaction mixture was concentrated, dissolved 3× in PhMe (0.250 L, 5.0 mL/g), and concentrated to remove excess AcOH. To the residue was added PhMe (0.200 L, 4.0 mL/g), and the mixture was warmed to 50°C for 2−5 h to effect crystallization. The mixture was cooled to 0°C and stirred for an additional 2−5 h. The solids were filtered, washed with PhMe (2 × 0.100 L, 2 × 2.0 mL/g), and dried in vacuo to provide the AcOH salt of aminopyridine.

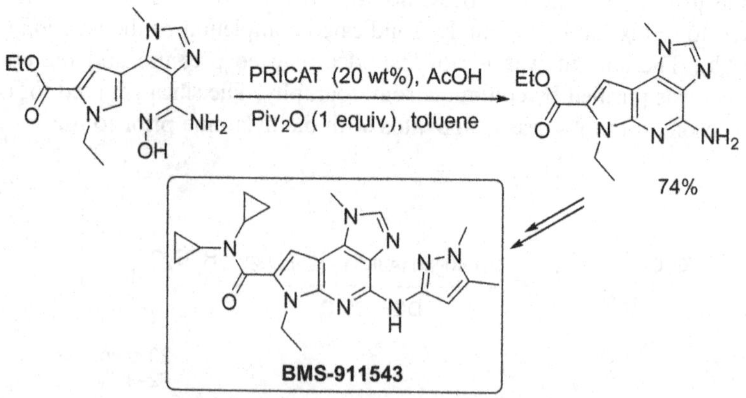

Scheme 2.4 Ni-catalyzed synthesis of JAK2 inhibitor BMS-911543.

Reddy's group reported a silver-catalyzed annulation of enynyl azides to pyridine derivatives in 2015 [8]. 3,6-Disubstituted pyridines were prepared in good yields through Ag-mediated aza-annulation of 2-en-4-ynyl azides which were derived from MBH-acetates of acetylenic aldehydes. 5-Iodo-3,6-Disubstituted pyridines can be prepared from enynyl azides having an electron-rich substituent on the alkyne functionality with I_2 as the promoter (Scheme 2.5).

Scheme 2.5

To a stirred solution of alkynyl azide (1 equiv.) in 1,2-dichloroethane (3.0 mL) was added $AgSbF_6$ (0.3 equiv.) and TFA (2 equiv.) at 80°C. After completion of the reaction, the mixture was quenched by saturated $NaHCO_3$ and stirred for 30 min. The mixture was extracted with CH_2Cl_2, organic layer was washed with H_2O, brine, dried over Na_2SO_4, and concentrated in vacuo. The residue was purified by column chromatography on silica gel (EtOAc: hexanes) to afford the corresponding product.

To a solution of azide (1 equiv.) in CH_2Cl_2 and $NaHCO_3$ (1 equiv.) was added at 0°C followed by the addition of iodine (5 equiv.), the solution was stirred at room temperature for given time. After completion of the reaction, the mixture was quenched with $Na_2S_2O_3$ solution and extracted with EtOAc, organic layer was washed with H_2O, brine, dried over Na_2SO_4 and concentrated in vacuo. The residue was purified by column chromatography on silica gel (EtOAc: hexanes) to afford the corresponding product.

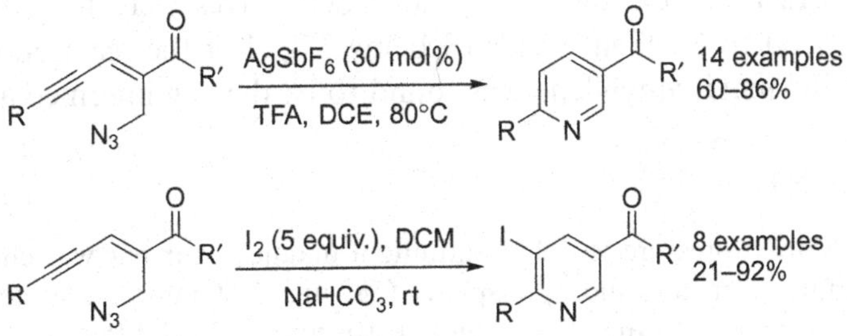

Scheme 2.5 Ag-mediated synthesis of pyridines from enynyl azides.

A Lewis acid mediated reaction of acyclic vinyl allenes and imines to produce tetrahydropyridines was developed by Palenzuela and coworkers in 2012 [9]. The cycloadducts can be transformed into polysubstituted pyridines, including bipyridines, by catalytic transfer hydrogenation using cyclohexene as hydrogen donor in good yield (Scheme 2.6). The reaction was extended to the aromatization of the bicyclic cycloadducts prepared in previous works from the reaction of semicyclic vinyl allenes and imines, yielding tetrahydropyridines. Metathesis was applied in heterocycle synthesis as well and can be further transformed into the corresponding pyridines [10].

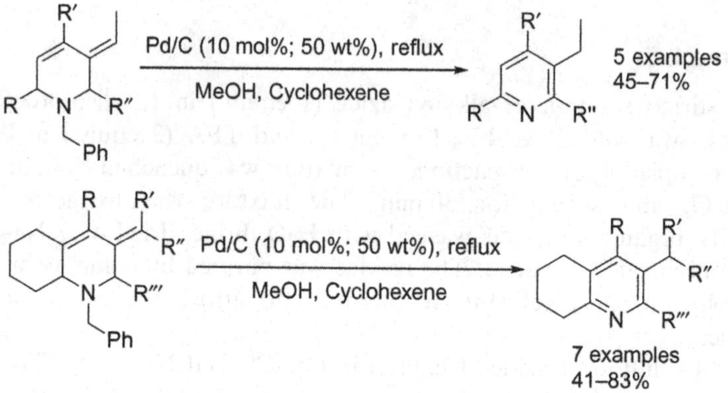

Scheme 2.6 Pd/C-catalyzed synthesis of pyridines from tetrahydropyridines.

Movassaghi and Hill developed a ruthenium-catalyzed cycloisomerization of 3-azadienynes to the corresponding pyridines [11]. The alkynyl imines were produced from a variety of N-vinyl and N-aryl amides by amide activation and nucleophilic addition of copper(I) (trimethylsilyl) acetylide sequence reaction. Then by Ru-catalyzed protodesilylation and cycloisomerization, the desired pyridine derivatives were formed selectively in good to excellent yields (Scheme 2.7). For the reaction mechanism, C-silyl metal vinylidene was found to be the key intermediate.

Scheme 2.7

An oven dried pressure vessel containing a magnetic stir bar was charged with ammonium hexafluorophosphate (587 mg, 3.60 mmol, 1.00 equiv.), CpRuCl(PPh$_3$)$_2$ (262 mg, 0.35 mmol, 0.10 equiv.), and SPhos (148 mg, 0.35 mmol, 0.10 equiv.) under a nitrogen atmosphere in a glove-box and the flask sealed and brought out of the glove-box. Imine (3.60 mmol, 1 equiv.) and toluene (18 mL) were subsequently added via syringe. The flask was flushed with argon, sealed, stirred, and placed in an oil bath at 105°C. After 19 h, the reaction vessel was allowed to cool to ambient temperature and the mixture was transferred to a recovery flask with a

20-mL portion of dichloromethane. This solution was concentrated under reduced pressure and the residue was purified by flash column chromatography on silica gel (15 → 50% EtOAc/Hexanes) to afford the quinoline as pure product.

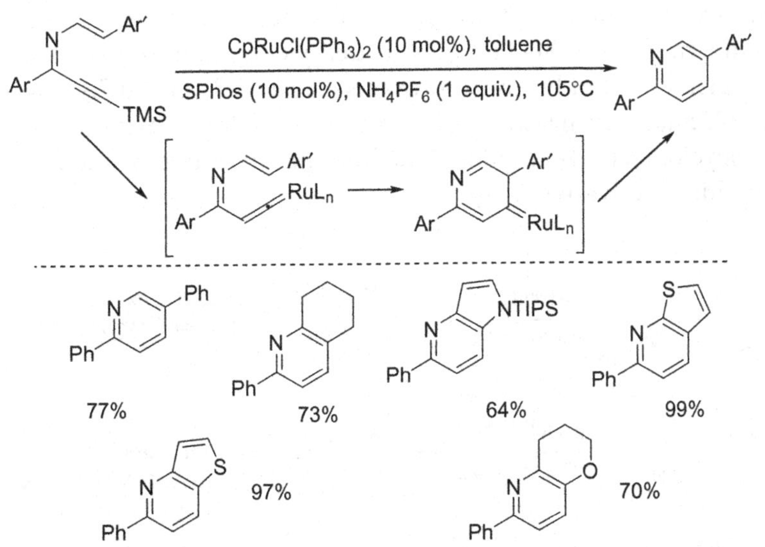

Scheme 2.7 Ru-catalyzed synthesis of pyridines from 3-azadienynes.

Hanzawa, Saito, and their coworkers developed a rhodium(I)-catalyzed intramolecular hetero-[4 + 2] cycloaddition of σ-alkynyl-vinyl oximes in 2007 [12]. By using [RhCl(cod)]₂ and AgSbF₆ as the catalyst system in hexafluoroisopropanol (HFIP), bicyclic pyridine derivatives were formed in good yields (Scheme 2.8).

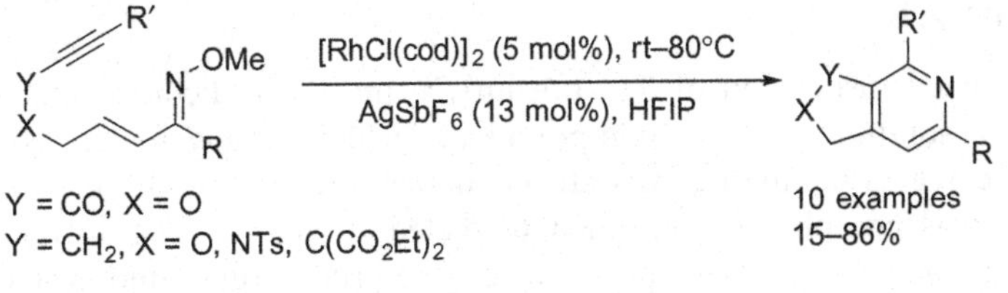

Y = CO, X = O
Y = CH₂, X = O, NTs, C(CO₂Et)₂

10 examples
15–86%

Scheme 2.8 Rh-catalyzed synthesis of pyridines from σ-alkynyl-vinyl oximes.

Zhang and Gao studied the silver(I)-catalyzed cyclization reaction of *ortho*-alkynylaryl aldehyde oxime derivatives [13]. A dramatic substituent effect was found in this transformation. When the oxime substituted with an alkyl group, this Ag(I)-catalyzed reaction gave isoquinolines good to excellent yields in dimethylacetamide at 110°C, in contrast, isoquinolin-1(2*H*)-ones were produced in moderate to high yields in dimethylformamide at room temperature when the oxime substituted with an acetyl group. In the case of substrates without an aromatic ring, Ph$_3$PAuOTf gave better yields than AgOTf (Scheme 2.9). Later on, Reddy and coworkers developed an AuCl$_3$-catalyzed intramolecular cyclization of aromatic *o*-alkynyloximes and 2-alkynylcycloalkene-1-carbaldoximes for the formation of isoquinoline and pyridine derivatives [14].

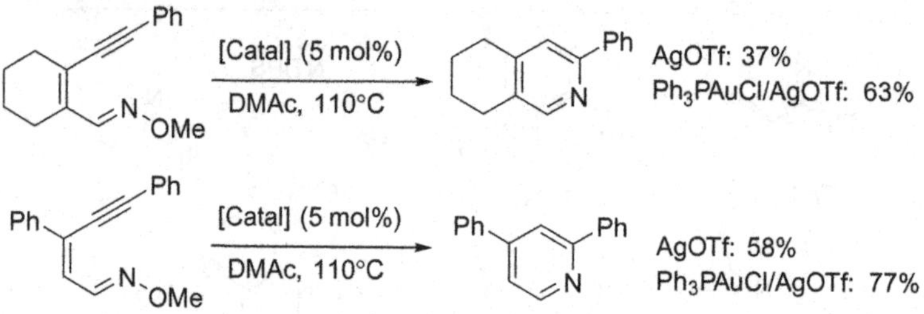

Scheme 2.9 Ag/Au-catalyzed synthesis of pyridines from oximes.

Park and coworkers developed a highly efficient procedure for the synthesis of 3-hydroxypyridines in 2012 [15]. With dirhodium complex Rh$_2$(tfacam)$_4$ (tfacam = trifluoroacetamide) as the catalyst and δ-diazo oxime ethers as substrates, based on carbenoid-mediated N−O bond insertion, a variety of pyridines were produced in 5−10 min in good to excellent yields. Recently, they reported their new achievements on transformation of α-diazo oxime ethers to pyridines with rhodium as the catalyst (Scheme 2.10).

Scheme 2.10

To a stirred suspension of Rh$_2$(tfacam)$_4$ (2 mol%) in dichloroethane was added a solution of diazo compound in dichloroethane under nitrogen, and the reaction mixture was stirred under reflux until the diazo compound was completely consumed (detected by TLC). The solvent was evaporated under reduced pressure to give crude compound which was purified by flash column chromatography using hexane : ethyl acetate (9:1) to give desired products.

A solution of diazo compound (0.2 mmol) and $Rh_2(OAc)_4$ (0.004 mmol, 2 mol%) in chlorobenzene (2.0 mL) was stirred at 130°C until the starting material was fully consumed. The reaction mixture was concentrated under reduced pressure to give the crude material which was purified by flash chromatography using hexane: ethyl acetate (9:1) to give desired products.

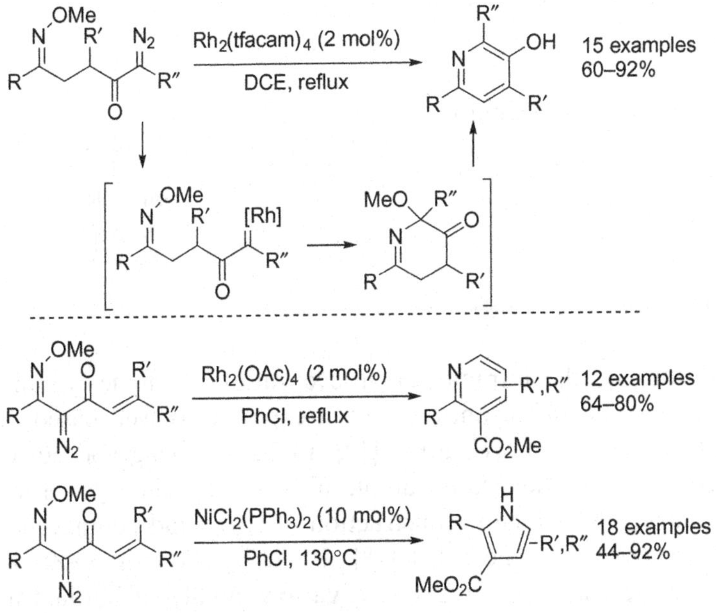

Scheme 2.10 *Rh-catalyzed synthesis of pyridines from diazo compounds.*

More recently, a visible-light-promoted cyclization of acyl oximes to pyridine derivatives via iminyl-radical was published by Yu, Zhang, and coworkers (Scheme 2.11) [16]. With fac-[Ir(ppy)$_3$] as the photoredox catalyst, the acyl oximes were converted by 1e$^-$ reduction into iminyl radical intermediates, which then underwent intramolecular homolytic aromatic substitution (HAS) to give the *N*-containing arenes. The reactions proceeded in high yield at room temperature with a broad range of substrates. This strategy of visible-light-induced iminyl-radical formation was successfully applied to a five-step concise synthesis of benzo[*c*]-phenanthridine alkaloids.

Scheme 2.11

A 10 mL round bottom flask was equipped with a rubber septum and magnetic stir bar and was charged with substrate (0.2 mmol, 1.0 equiv.), fac-Ir(ppy)$_3$ (0.002 mmol, 0.01 equiv.). The flask was evacuated and back-filled with N_2 three times. DMF (4.0 mL, 0.05 M) was then added with syringe under N_2. The mixture was then irradiated by a 5W white LEDs strip. After the reaction was complete (as judged by TLC analysis), the mixture was concentrated under vacuum to remove DMF. The residue was then purified by flash chromatography on silica gel (EtOAc: hexane = 1:10) to afford the pure product.

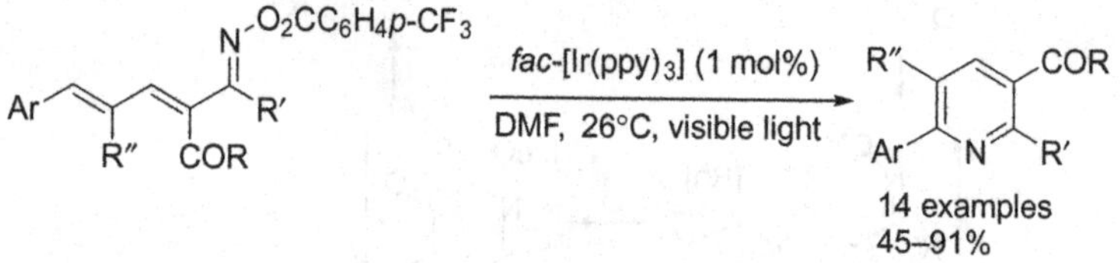

Scheme 2.11 Light-promoted synthesis of pyridines from acyl oximes.

Additionally, the intramolecular cyclization of diyne-cyanohydrin (two alkynes and one cyano) has been explored as well based on the concept of (2 + 2 + 2) cyclization [17]. In 2007, Cheng and coworkers reported a cobalt-catalyzed intramolecular [2 + 2 + 2] cocyclotrimerization of nitrilediynes for the construction of tetra- and pentacyclic pyridine derivatives (Scheme 2.12) [18]. By using CoI$_2$(dppe)/Zn as the catalytic system at 80°C in CH$_3$CN, various highly substituted nitrilediynes were transformed into the corresponding products in good to excellent yields (Scheme 2.12). Afterwards, several modified cobalt catalyst-based systems were also developed [19–21].

Scheme 2.12

A 25 mL round-bottomed side-arm flask containing CoI$_2$(dppe) (0.005 mmol), Zn (2.75 mmol) was evacuated and purged with nitrogen gas three times. To the flask were then added nitrilediyne (1.00 mmol) and CH$_3$CN (3.0 mL) via syringes. The reaction mixture was allowed to stir at 80°C for 12 h. At the end of the reaction, the reaction mixture was diluted with CH$_2$Cl$_2$, filtered through Celite and silica gel, and the filtrate was concentrated. The crude residue was purified through a silica gel column using hexanes and ethyl acetate as eluent to give pure products.

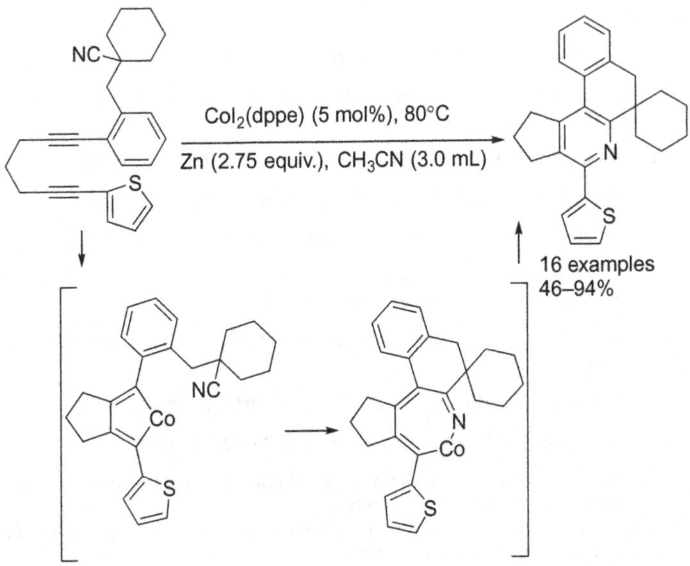

Scheme 2.12 Cobalt-catalyzed synthesis of pyridines from nitrilediynes.

Rhodium catalysts were applied in this type of transformation as well. Notably, by the addition of enantioenriched C2-symmetric spiro-bipyridine ligands, the reaction proceeded in an enantioselective manner (Scheme 2.13a) [22]. More recently, cyano-yne-allenes were found to be suitable substrates as well [23]. 2,6-Naphthyridine derivatives can be produced selectively here (Scheme 2.13b).

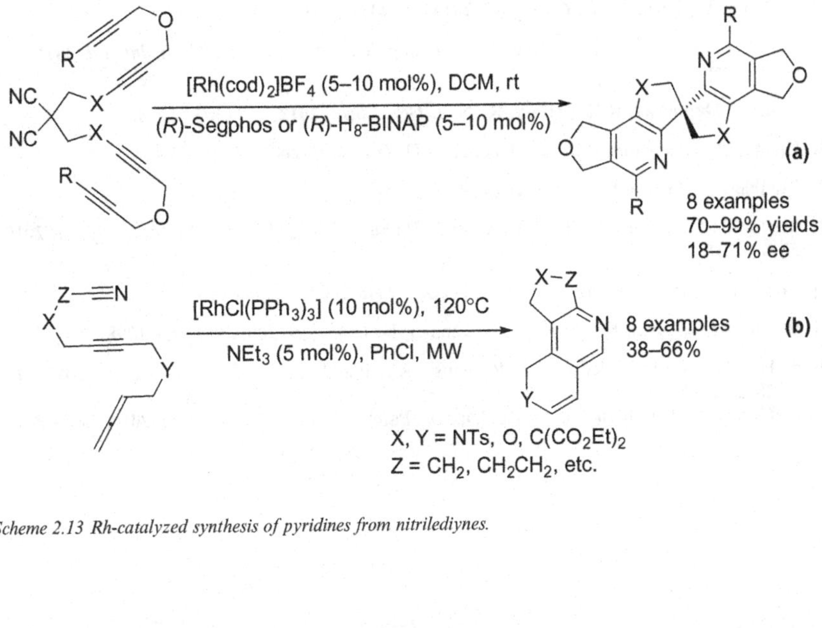

Scheme 2.13 Rh-catalyzed synthesis of pyridines from nitrilediynes.

REFERENCES

[1] Cacchi, S.; Fabrizi, G.; Filisti, E. *Org. Lett.* **2008**, *10*, 2629–2632.

[2] Bagley, M. C.; Brace, C.; Dale, J. W.; Ohnesorge, M.; Phillips, N. G.; Xiong, X., et al. *J. Chem. Soc., Perkin Trans.* **2002**, *1*, 1663–1671.

[3] Ohashi, M.; Kamachi, H.; Kakisawa, H.; Stork, G. *J. Am. Chem. Soc.* **1967**, *89*, 5460–5461.

[4] Kim, K. H.; Kim, S. H.; Lee, H. J.; Kim, J. N. *Adv. Synth. Catal.* **2013**, *355*, 1977–1983.

[5] Prechter, A.; Henrion, G.; dit Bel, P. F.; Gagosz, F. *Angew. Chem. Int. Ed.* **2014**, *53*, 4959–4963.

[6] Jin, L.; Wu, Y.; Zhao, X. *J. Org. Chem.* **2015**, *80*, 3547–3555.

[7] Fitzgerald, M. A.; Soltani, O.; Wei, C.; Skliar, D.; Zheng, B.; Li, J., et al. *J. Org. Chem.* **2015**, *80*, 6001–6011.

[8] Reddy, C. R.; Panda, S. A.; Reddy, M. D. *Org. Lett.* **2015**, *17*, 896–899.

[9] Regás, D.; Afonso, M. M.; Palenzuela, J. A. *Tetrahedron* **2012**, *68*, 9345–9349.

[10] a. Donohoe, T. J.; Bower, J. F.; Basutto, J. A.; Fishlock, L. P.; Procopiou, P. A.; Callens, C. K. A. *Tetrahedron* **2009**, *65*, 8969–8980.
b. Donohoe, T. J.; Bower, J. F.; Baker, D. B.; Basutto, J. A.; Chan, L. K. M.; Gallagher, P. *Chem. Commun.* **2011**, *47*, 10611–10613.
c. Donohoe, T. J.; Basutto, J. A.; Bower, J. F.; Rathi, A. *Org. Lett.* **2011**, *13*, 1036–1039.
d. Yoshida, K.; Kawagoe, F.; Hayashi, K.; Horiuchi, S.; Imamoto, T.; Yanagisawa, A. *Org. Lett.* **2009**, *11*, 515–518.

[11] a. Movassaghi, M.; Hill, M. D. *J. Am. Chem. Soc.* **2006**, *128*, 4592–4593.
b. Hill, M. D.; Movassaghi, M. *Synthesis* **2007**, 1115–1119.

[12] Saito, A.; Hironaga, M.; Oda, S.; Hanzawa, Y. *Tetrahedron Lett.* **2007**, *48*, 6852–6855.

[13] Gao, H.; Zhang, J. *Adv. Synth. Catal.* **2009**, *351*, 85–88.

[14] Subbarao, K. P. V.; Reddy, G. R.; Muralikrishna, A.; Reddy, K. V. *J. Heterocycl. Chem.* **2014**, *51*, 1045–1050.

[15] a. Qi, X.; Dai, L.; Park, C.-M. *Chem. Commun.* **2012**, *48*, 11244–11246.
b. Jiang, Y.; Park, C.-M. *Chem. Sci.* **2014**, *5*, 2347–2351.

[16] Jiang, H.; An, X.; Tong, K.; Zheng, T.; Zhang, Y.; Yu, S. *Angew. Chem. Int. Ed.* **2015**, *54*, 4055–4059.

[17] Lan, Y.; Danheiser, R. L.; Houk, K. N. *J. Org. Chem.* **2012**, *77*, 1533–1538.

[18] Chang, H.-T.; Jeganmohan, M.; Cheng, C.-H. *Org. Lett.* **2007**, *9*, 505–508.

[19] Meißner, A.; Groth, U. *Synlett* **2010**, 1051–1054.

[20] Miclo, Y.; Garcia, P.; Evanno, Y.; George, P.; Sevrin, M.; Malacria, M., et al. *Synlett* **2010**, 2314–2318.

[21] Zhou, Y.; Porco, J. A., Jr.; Snyder, J. K. *Org. Lett.* **2007**, *9*, 393–396.

[22] Wada, A.; Noguchi, K.; Hirano, M.; Tanaka, K. *Org. Lett.* **2007**, *9*, 1295–1298.

[23] a. Haraburda, E.; Lledo, A.; Roglans, A.; Pla-Quintana, A. *Org. Lett.* **2015**, *17*, 2882–2885.
b. Garcia, L.; Pla-Quintana, A.; Roglans, A.; Parella, T. *Eur. J. Org. Chem.* **2010**, 3407–3415.

CHAPTER 3

Synthesized by Intermolecular Cyclizations

3.1 SYNTHESIZED BY [2 + 2 + 2] CYCLIZATION REACTIONS

Formal [2 + 2 + 2] cyclization reactions are the most explored transition metal-catalyzed procedures for pyridines synthesis. In this section, the main achievements on this topic will be discussed based on the catalyst systems applied. Takahashi's group reported a methodology for the synthesis of pyridines, pyridones, and iminopyridines from two different alkynes and a nitrile *via* azazirconacycles [1]. In the presence of 1 equiv. of $NiCl_2(PPh_3)_2$, through azazirconacyclopentadienes intermediate only single products of pyridine derivatives from two different symmetrical alkynes and a nitrile can be achieved. In the same period, Sato and their coworkers developed a titanium(II) alkoxide-mediated pyridine synthesis from two different unsymmetrical acetylenes and a nitrile [2]. Notably, tris(acetonitrile) complex $[Fe(C_5Me_5)(CH_3CN)_3][PF_6]$-mediated [3] and cobalt-mediated [4] versions were reported as well. In 1977, Vollhardt and Naiman reported their work on $CpCo(CO)_2$ catalyzed synthesis of pyridines [5]. By refluxing in *o*-xylene, 47% of the desired pyridine can be produced from 1,7-octadiyne and cyanoacetic ester.

Heller and coworkers performed cobalt-catalyzed photochemical synthesis of pyridines [6]. Ethyne, propyne, butyne, dimethylbutyne, benzo-, aceto-, trimethoxybenzonitrile, *n*-butylcyanoacetate, *tert*-butylcyanide and etc. were all tested as substrates. And cobalt complexes YCo(cod) were applied as the catalysts (Y = η^5-cyclopentadienyl, η^5-indenyl, η^5-tetraphenylcyclopentadienyl, η^5-acetylcyclopentadienyl, η^3- cyclooctenyl, η^6-1-phenylborinato, and so on). Very high chemoselectivities were obtained at low alkyne concentrations. The regioselectivity can be advantageously influenced by the properties of the ligands in the precatalyst and by the application of bulky nitriles and alkynes. Both artificial light and sunlight can also be employed as the radiation sources without loss of selectivity. Remarkably, by using chiral cobalt(I) complexes as the catalysts, the reaction can be realized in an asymmetric manner. Hagiwara, Hoshi, and coworkers applied this concept in the synthesis of chiral 2,2'-bis(pyridin-2-yl)-1,1'-binaphthyl ligands from enantio

Transition Metal-Catalyzed Pyridine Synthesis. DOI: http://dx.doi.org/10.1016/B978-0-12-809379-5.00003-2

pure 2,2′-dicyano-1,1′-binaphthyl with 1,6-heptadiyne, 1,7-octadiyne, 1,8-nonadiyne, and 2,8-decadiyne [7]. Microwave heating was applied in [2 + 2 + 2] cocyclization as well [8]. Kotora and coworkers reported a methodology for the preparation of a series of (pyridin-2-yl)purine derivatives in 2008. Good yields of pyridines were produced from 6-(diynyl)purines and nitriles in the presence of a stoichiometric or catalytic amount of [CpCo(CO)₂] enhanced by microwave. Alkyl, aryl, and heteroaryl cyanides are all suitable substrates for this transformation.

Deiters and coworkers developed a solid-supported version of [2 + 2 + 2] cocyclization of alkynes and nitriles for pyridines synthesis [9]. By using CpCo(CO)$_2$ (20 mol%) as the precatalyst and tetramethylammonium oxide (TMAO) as the catalyst activator at 80°C, good yields of the desired pyridines can be achieved. In the procedure, one of the alkyne was immobilized on the resin and cleaved under mild conditions using 1% TFA in DCM. More recently, Hapke and Thiel supported air-stable CpCo(I)-complex [CpCo(trans-MeO$_2$CCH CHCO$_2$Me){P(OEt)$_3$}] [10] in order to enhance its recyclability [11]. By covalently linking this highly active catalyst on silica-bound analogue support, pyridine derivatives can be produced in good yields by cyclotrimerization of a diyne and a nitrile.

A systematic study on the use of [CpCo(CO)(dmfu)] (dmfu = dimethyl fumarate) [12] as a precatalyst for the cocyclization of alkynes and nitriles was published in 2011 [13]. By this catalyst, the incorporation of electron-deficient nitriles into the pyridine core was realized. 3- or 4-Aminopyridines can be produced regioselectively by modifying the substitution pattern at the yne-ynamide. Based on DFT computations, the author suggested that 3-aminopyridines are formed by formal [4 + 2] cycloaddition between the nitrile and the intermediate cobaltacyclopentadiene, whereas 4-aminopyridines arise from an insertion pathway. This catalytic system was applied in the synthesis of bicyclic 3- or 4-aminopyridines from yne-ynamides and nitriles (Scheme 3.1) [14].

Scheme 3.1

To a solution of the starting ynamide in toluene were added the nitrile and CpCo(CO)dmfu under inert atmosphere. After being refluxed for 15 h and allowed to cool to rt, the mixture was purified by flash chromatography eluting first with petroleum ether and then (unless otherwise noted) with petroleum ether/ethyl acetate (9:1).

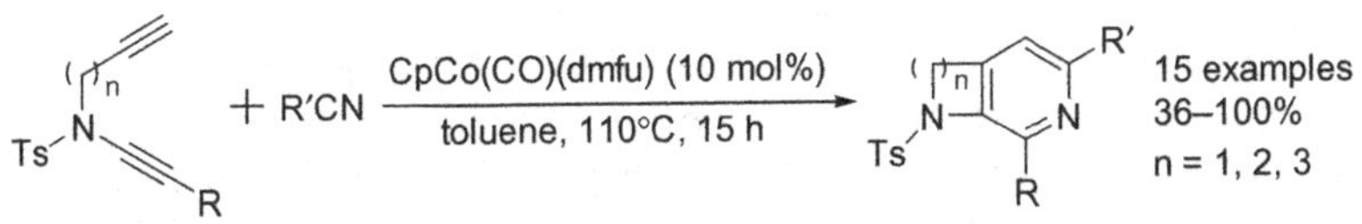

Scheme 3.1 Co-catalyzed synthesis of 3- or 4-aminopyridines.

Saá and coworkers reported a one-step procedure for the synthesis of spiropyridines in 1999 [15]. The desired pyridines can be obtained in moderate yields by Co(I)-catalyzed double cocyclization between bis-alkynenitriles and alkynes (Scheme 3.2). The products can be applied as $C2$-symmetric chiral ligands and a diastereomeric mixture of tetracoordinated [Cu^{I}L$_2$] complex with 7,7′-spiropyridine was prepared as well.

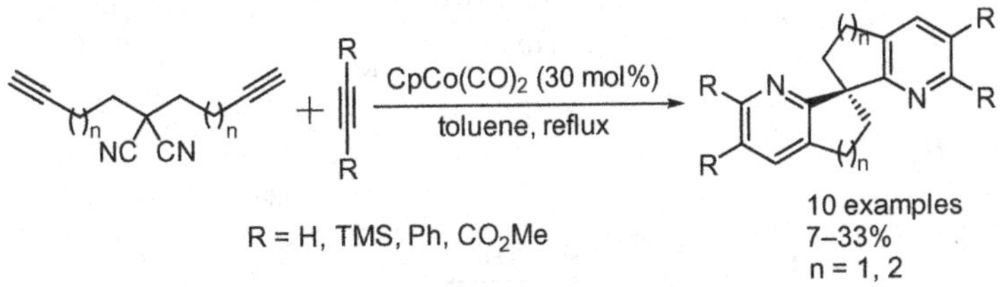

Scheme 3.2 Co-catalyzed synthesis of spiropyridines.

In 2008, Schreiber's group studied the cycloaddition of silyl-tethered diynes with nitriles [16]. The tether provided high regioselectivity, and no exogenous irradiation required while using THF as the solvent with catalytic amount of CpCo(CO)$_2$. Moderate to excellent yields of the desired products can be achieved (Scheme 3.3). One of the resulting bicyclic and monocyclic (desilylated) pyridines was identified as an inhibitor of neuregulin-induced neurite outgrowth (EC$_{50}$ = 0.30 μM) in a screen that probes a pathway likely to be involved in breast cancers and schizophrenia.

Scheme 3.3

Diisopropyl(pent-1-yn-3-yloxy)-(phenylethynyl)silane (10.0 mg, 0.033 mmol) was placed into an oven dried sealed tube and dissolved in degassed THF (0.67 mL). Nitriles (0.05 mmol, 1.5 equiv.) were added to the stirring solution. After complete dissolution, a solution of cyclopentadienylcobalt (I) dicarbonyl (1.8 mg, 0.010 mmol, 30 mol%) in degassed xylenes (50 μL) was introduced by syringe, giving a pale yellow solution that was submerged into an oil bath preheated to 140°C. After 24 h, the dark brown solution was cooled to ambient temperature and loaded onto a 4 g silica plug. Filtration was performed using an Isco Combiflash system using a gradient solvent commencing with hexanes and ending with 1/1 hexanes/ethyl acetate (total volume of approximately 40 mL), which effectively removed insoluble cobalt byproducts. Pooled fractions were concentrated in vacuo and assayed for conversion by ^1H NMR. Purification was performed by silica gel chromatography using an Isco Combiflash 12 g column with 20:1 hexanes/EtOAc as eluant, providing the corresponding pyridines.

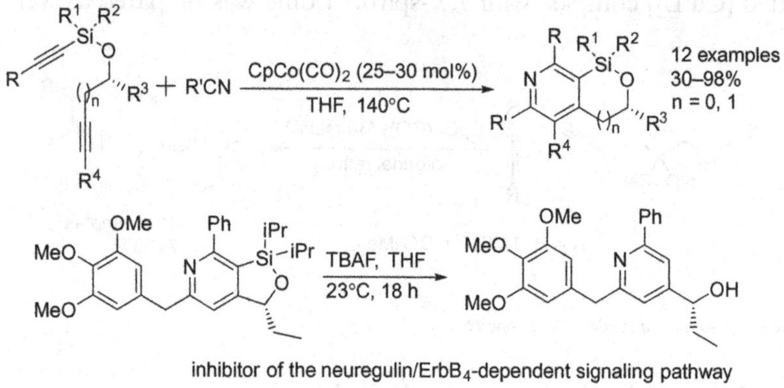
CpCo(CO)2 (25–30 mol%)
THF, 140°C
R'CN
12 examples
30–98%
n = 0, 1
TBAF, THF
23°C, 18 h
inhibitor of the neuregulin/ErbB4-dependent signaling pathway

Scheme 3.3 Co-catalyzed synthesis of silylated pyridines.

In the same year, Deiters and coworkers developed an approach for the construction of anthracenes and 2-azaanthracenes via [2 + 2 + 2] cyclotrimerization [17]. Good yields of the products can be isolated and tests in fluorescence emission and the fluorescent labeling of mammalian cells were performed as well (Scheme 3.4). These compounds have unique photochemical and biological properties and can act as environmentally sensitive dyes, metal sensors, pH sensors, and cellular stains.

Scheme 3.4

The diyne (20 mg, 0.13 mmol), nitrile (1.3 mmol), dry toluene (4 mL), and $CpCo(CO)_2$ (1.56 μL, 0.013 mmol) were added to a flame dried microwave vial equipped with a stir bar. The vial was flushed with nitrogen, capped with a microwave vial septum and irradiated for 20 min in a

CEM Discover microwave synthesizer at 300 W. After cooling, the reaction was purified by silica gel chromatography, eluting with hexanes/EtOAc. Due to compound instability, the products were only characterized via NMR and directly subjected to the next reaction without HRMS measurement.

The precursors, DDQ (1.2 equiv.), and dry toluene (2 mL) were added to a flame dried microwave vial equipped with a stir bar. The vial was flushed with nitrogen, capped with a microwave vial septum and irradiated for 5 min in a CEM Discover microwave synthesizer at 300 W.

The azaanthracenes (0.02 mmol) and MeI (13 μL, 0.2 mmol) were added to a flame-dried vial under nitrogen atmosphere. The reaction mixture was heated at 60°C overnight. After cooling to room temperature, the solvent was removed under reduced pressure to yield the desired products in quantitative yield.

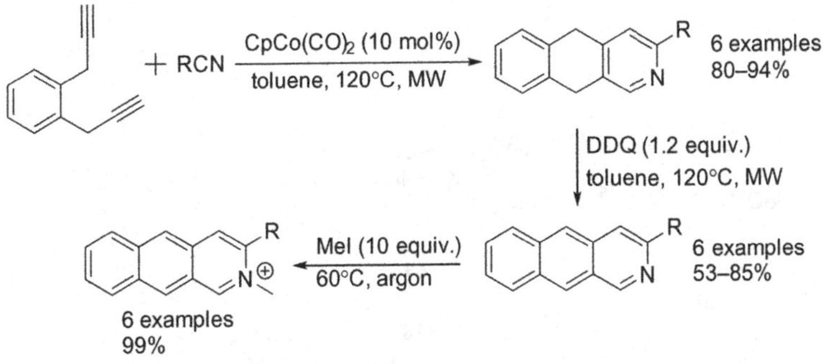

Scheme 3.4 Co-catalyzed synthesis of azaanthracenes.

A highly practical method for the catalytic formation of substituted pyridines from a variety of unactivated nitriles and α,σ-diynes was developed by Okamoto and coworkers in 2007 [18]. With 5 mol% of dppe/CoCl$_2$·6H$_2$O as the catalyst in the presence of Zn powder (10 mol %), pyridines were produced in good yields with high functional compatibility and regioselectivity at rt to 50°C (Scheme 3.5). This system was also applied in the synthesis of polymers by polymerization of diyne-nitrile monomers [19]. With 2-(2-(hept-2-yn-1-yl)non-4-yn-1-yl) malononitrile or 2,2-di(but-2-yn-1-yl)malononitrile as the substrate, the corresponding polymers can be selectively produced.

Scheme 3.5

To a mixture of diyne (1.0 mmol), nitrile (1.5 ~ 20 mmol) and zinc powder (6.5 mg, 0.10 mmol) in NMP (0.6 mL) was added a solution of $CoCl_2 \cdot 6H_2O$ (11.9 mg, 0.05 mmol) and dppe (23.9 mg, 0.06 mmol) in NMP (0.4 mL) (see note below). The resulting mixture was stirred at ambient temperature. After completion of the reaction checked by TLC analysis, Et_2O (5 mL) was added and the mixture was filtered through a pad of Celite with ether (5 mL). The filtrate was concentrated in vacuo and chromatographed on silica gel to give the corresponding substituted pyridine.

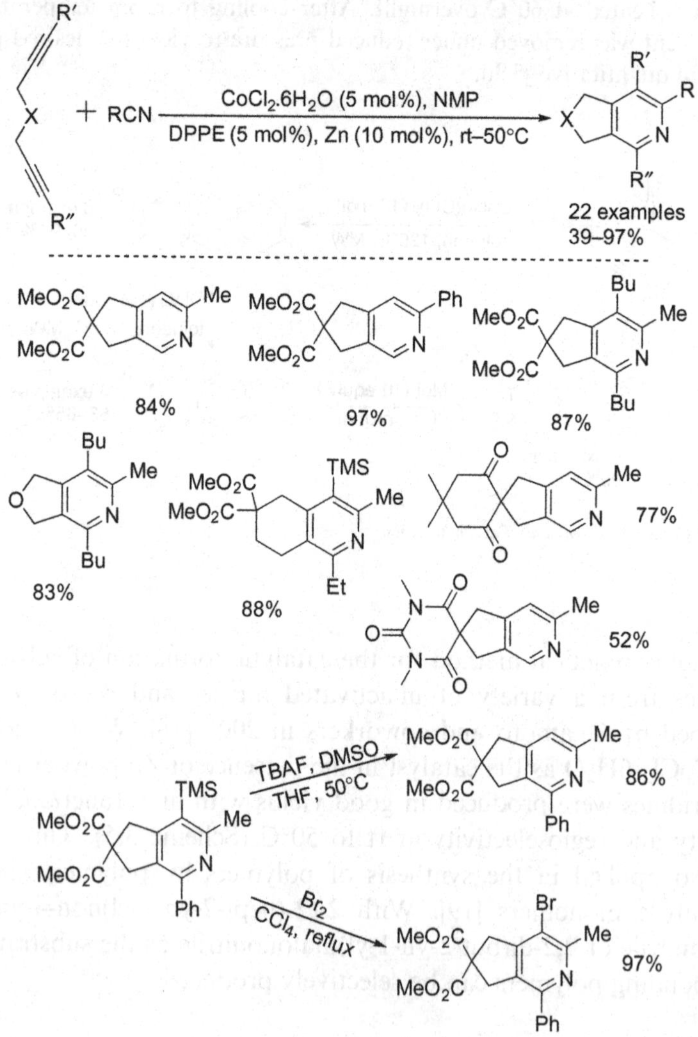

Scheme 3.5 CoCl₂-catalyzed synthesis of pyridines.

Schmalz and Nicolaus applied Co-catalyzed cyclization in the synthesis of pyridine incorporated allocolchicine analogues [20]. Started from readily accessible 3,4,5-trimethoxybenzaldehyde or 3,4,5-trimethoxyacetophenone, the (racemic) target compounds were regioselectively obtained in 15–20% overall yield (five steps; Scheme 3.6).

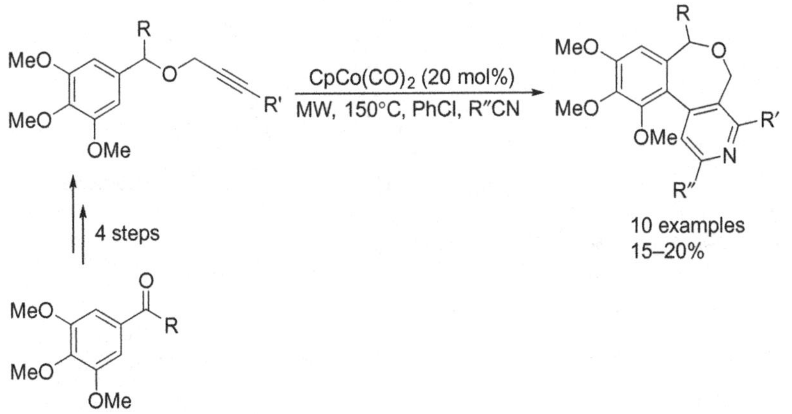

Scheme 3.6 Co-catalyzed synthesis of allocolchicine analogues.

Saá, Vollhardt, and their coworkers applied cobalt catalyst in the synthesis of Lysergene and LSD [21]. Lysergene is a naturally occurring clavine. LSD is also known as lysergide (INN), is a psychedelic drug of the ergoline family, well known for its psychological effects and mainly used as an entheogen and recreational drug. By using 4-ethynyl-3-indoleacetonitriles and alkynes as the substrates in the presence of cobalt catalyst, the core structures can be prepared (Scheme 3.7).

Complanadine A is a member of the large family of lycopodium alkaloids, it's naturally from the club moss Lycopodium complanatum. It has been reported to induce the secretion of neurotrophic factors from 1321N1 cells, promoting the differentiation of PC-12 cells. Maturely, complanadine A is likely synthesized through the union of two molecules of lycodine. In 2010, Siegel and coworkers applied Co(I)-mediated [2 + 2 + 2] in the total synthesis of Complanadine A (Scheme 3.8) [22].

Saá, Castedo, and coworker applied cobalt(I)-catalyzed [2 + 2 + 2] cycloaddition in the synthesis of bipyridines and terpyridines in 1997 [23]. By using 5-hexynenitril, 6-heptynenitrile 2-ethynylpyridines, and 2,6-bis[(trimethylsilyl)ethynyl]pyridines as the substrates, the desired

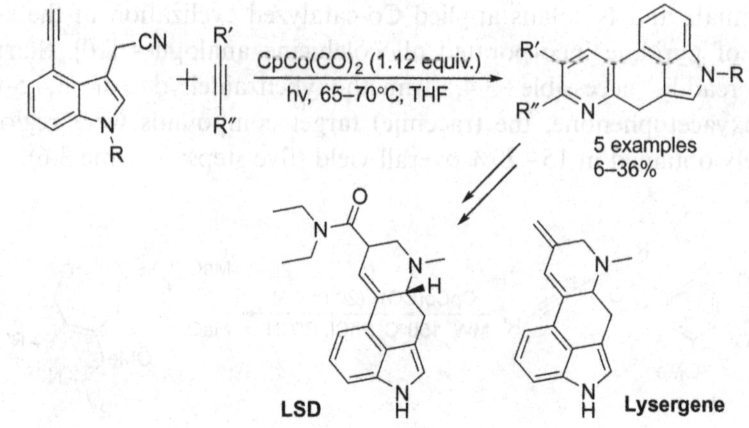

Scheme 3.7 *Co-mediated synthesis of Lysergene and LSD cores.*

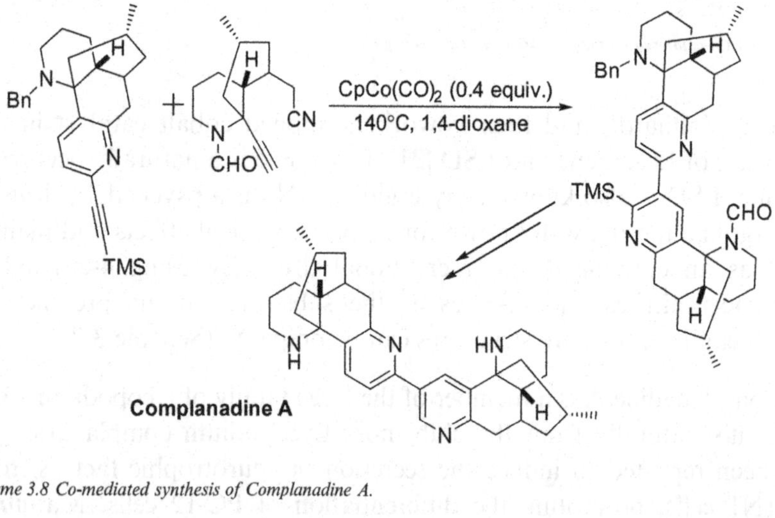

Scheme 3.8 *Co-mediated synthesis of Complanadine A.*

pyridine derivatives can be produced in moderate to good yields (Scheme 3.9a). Later on in 1998, they developed a new, one-step method for the synthesis of annelated symmetric 3,3′-substituted 2,2′-bipyridines from acyclic precursors [24]. By using CpCo(CO)$_2$ as the catalyst, with 5-hexynenitrile and 1,3-diynes as the starting materials, bipyridines can be selectively produced (Scheme 3.9b).

Scheme 3.9b

A solution of diyne (0.72 mmol), 5-hexynenitrile (0.2 g, 2.17 mmol), and [CpCo(CO)$_2$] (0.22 mmol) in toluene (10 mL) was irradiated for 1 h under Ar in a round-bottomed flask equipped with a reflux condenser. The reaction vessel was irradiated with a Philips PF 808 300 W tungsten slide projector lamp placed ca. 5 cm from the center of the flask and operated at 225 W. The volatile components were removed under vacuum and the residue was chromatographed on silica gel (100:0 to 90:10 EtOAc/MeOH).

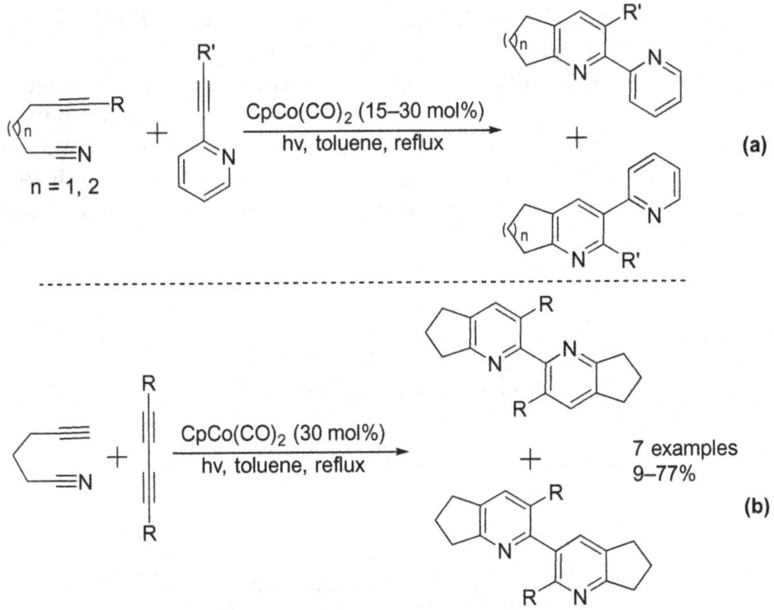

Scheme 3.9 Co-catalyzed synthesis of bipyridines.

Okamoto and coworkers reported a cobalt-catalyzed cycloaddition reaction of nitriles and α,σ-diyne to the synthesis of substituted 2,2′-bipyridines and 2,2′:6′,2″-terpyridines in 2008 [25]. A variety of substituted 2,2′-bipyridines were synthesized by a 1,2-bis(diphenylphosphino)-ethane (DPPE)/cobalt chloride hexahydrate (CoCl$_2 \cdot$ 6H$_2$O)/zinc-catalyzed [2 + 2 + 2] cycloaddition reaction from the corresponding substrates with excellent regioselectivity (Scheme 3.10). More specifically, symmetrical and unsymmetrical 1,6-diynes and 2-cyanopyridine reacted in the presence of 5 mol% of dppe, 5 mol% of

$CoCl_2 \cdot 6H_2O$ and 10 mol% of zinc powder to provide the corresponding 2,2′-bipyridines. Under identical reaction conditions, 1-(2-pyridyl)-
1,6-diynes and nitriles reacted smoothly to produce 2,2′-bipyridines in
good yield. 2,2′-Bipyridines were also obtained by the double
[2 + 2 + 2] cycloaddition reaction of 1,6,8,13-tetraynes with nitriles.
Similarly, 2,2′:6′,2″-terpyridines were synthesized from 1-(2-pyridyl)-
1,6-diyne and 2-cyanopyridine.

Scheme 3.10

To a stirred mixture of zinc powder (3.5 mg, 0.05 mmol), diyne or tetra-
yne (0.5 mmol) and nitrile (1.5−80 equiv.) in NMP (1 mL) was added a
solution of $CoCl_2 \cdot 6H_2O$ (6 mg, 0.025 mmol) and DPPE (12 mg,
0.03 mmol) in NMP (1 mL) at room temperature. The mixture was then
stirred at room temperature or at 50°C. The reaction progress was monitored by TLC analysis. After completion of the reaction, a small portion
of EtOAc or ether was added and the mixture was passed through a pad
of Celite with EtOAc or ether. The filtrate was concentrated to dryness
and the residue was chromatographed on silica gel using hexane/AcOEt
to give the corresponding bipyridine derivative.

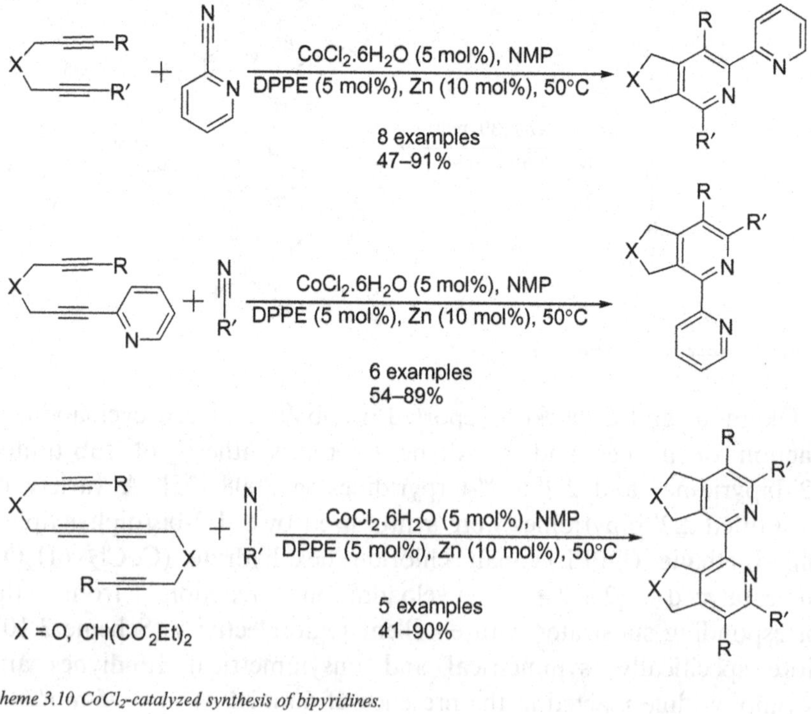

Scheme 3.10 CoCl₂-catalyzed synthesis of bipyridines.

One direction for the development in this area is looking for more active and stable complex. In 2011, Hapke and coworkers prepared $CpCo(H_2C = CHSiMe_3)_2$ complex and was found active in $[2 + 2 + 2]$ cycloaddition reactions [26]. The other direction is the exploration of applications. In 2005, Maryanoff and coworkers applied $[2 + 2 + 2]$ cycloaddition reaction in the preparation of macrocycles [27]. In the presence of $CpCo(CO)_2$, pyridine containing macrocycles were prepared from long-chain α,σ-diyne and nitriles, cyanamides, or isocyanates (Scheme 3.11). The regioselectivity of these reactions was affected by the length and type of linker unit between the alkyne groups, as well as by certain stereoelectronic factors.

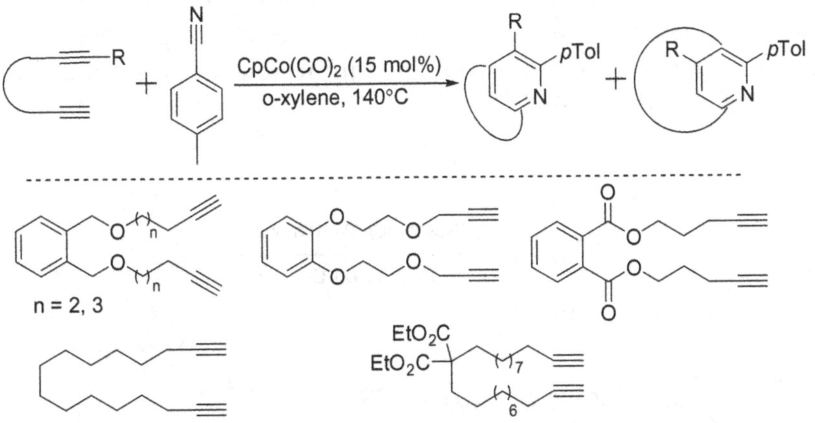

Scheme 3.11 Co-catalyzed synthesis of pyridine containing macrocycles.

In addition to cobalt catalysts, ruthenium catalysts were applied in $[2 + 2 + 2]$ cycloaddition reactions as well. In 2001, Itoh and coworkers reported a ruthenium-catalyzed cyclization of 1,6-diynes with dicyanides to produce the desired bicyclic pyridines in good yields [28]. By applying $Cp*Ru(cod)Cl$ ($Cp* = $ pentamethylcyclopentadienyl) as the catalyst, good yields of the products can be achieved (Scheme 3.12). Meanwhile, they explored the catalyst system in cyclization of 1,6-diynes with electron deficient nitriles as well. The desired bicyclic pyridines can be isolated in moderate to high yields [29]. Later on, in 2005, Yamamoto and coworkers performed systematic studies on this transformation and DFT calculations [30]. In the case of cyclization of 1,6-diynes with α-halonitriles to give 2-haloalkylpyridines, the effect of halide was explored [31]. In the presence of 2−5 mol% Cp*Ru(cod)Cl, various 1,6-diynes reacted with α-monohalo- and α,α-dihalonitriles at ambient temperature to afford 2-haloalkylpyridines in 42−93% isolated yields. Acetonitrile, N,N-dimethylaminoacetonitrile, phenylthioacetonitrile, and methyl cyanoacetate failed as nitrile substrates. The cycloaddition of unsymmetrical diynes bearing a substituent on one alkyne terminal gave 2,3,4,6-substituted pyridines exclusively. In their further studies, they found the nitrile components with methoxy and methylthio groups can behave as assisting groups, whereas nitrogen functional groups, such as pyridyl and amino groups, proved to be totally ineffective. Additionally, a trimethylsilylated alkynyl group showed a similar efficiency as the cyano group, although alkenyl and ester functionalities are incompetent. These results suggest that multiple lone pairs or π-bonds play a critical role in this ruthenium-catalyzed transformation [32].

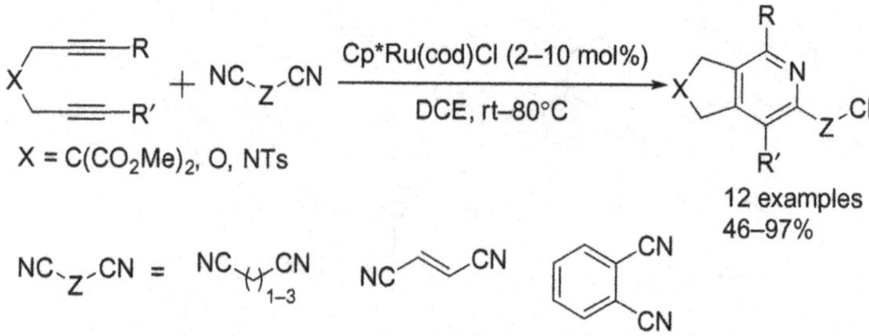

Scheme 3.12 Ru-catalyzed synthesis of pyridines.

In 2003, Saá and coworkers performed a comprehensive study on cationic [Cp*Ru(CH$_3$CN)$_3$]PF$_6$ complex-catalyzed [2 + 2 + 2] cycloaddition of 1,6-diynes to α,σ-dinitriles or electron-deficient nitriles (Scheme 3.13) [33]. The reaction with asymmetric electron-deficient alkynes could give the corresponding 2,3,6-trisubstituted pyridines in good yield. Based on their studies, they propose that the reactions with dinitriles seem likely to proceed via ruthenacyclopentadiene intermediates and the reactions with electron-poor nitriles via azaruthenacyclopentadienes.

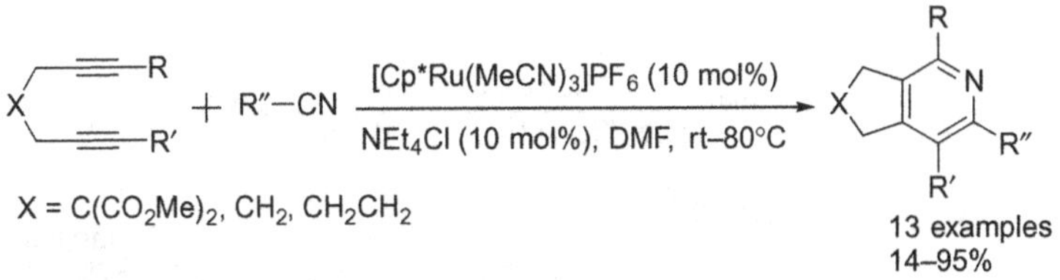

R

X ⟨≡—R / ≡—R′⟩ $+$ R″—CN $\xrightarrow{\text{[Cp*Ru(MeCN)}_3\text{]PF}_6\ (10\ \text{mol}\%)}{\text{NEt}_4\text{Cl (10 mol\%), DMF, rt–80°C}}$

X = C(CO₂Me)₂, CH₂, CH₂CH₂

Ring product with N, R, R″, R′

13 examples
14–95%

Scheme 3.13 [Cp Ru(CH₃CN)₃]PF₆-catalyzed synthesis of pyridines.*

Hoveyda-Grubbs' catalyst has been applied in [2 + 2 + 2] cyclotrimerizations of diynes with nitriles as well [34]. In the presence of 5 mol % of catalyst in DCE at 90°C, excellent yields can be obtained with activated nitriles and low to moderate yields with nonactivated nitriles. Both terminal and internal alkynes are applicable as substrates.

In 2012, Wan and coworkers reported a ruthenium-catalyzed process with pure water as the solvent (Scheme 3.14) [35]. Highly functionalized pyridines were produced selectively from both hydrophobic and hydrophilic diynes with tppts (tris(*m*-sulfonatophenyl) phosphine) as the water soluble ligand.

Scheme 3.14

A mixture of 5 mol% Cp*Ru(COD)Cl (4.7 mg, 0.025 mmol), 20 mol% tppts (28.4 mg, 0.1 mmol), and pure water (2 mL, oxygen-free) was stirred at 100°C for 30 min, resulting in the formation of an orange-red aqueous phase. After the solution was cooled to 25°C and stirred for 30 min, the nitrile was added, and the reaction mixture was stirred at 25°C for an additional 5 min. After the addition of the nitrile an emulsion formed. Diyne was then added, and the mixture was kept stirring for 5 min at room temperature. The reaction mixture was then stirred at 50°C for 48 h. After cooling to room temperature, the aqueous reaction mixture was extracted with CH₂Cl₂ (2 × 5 mL). The combined organic phase was washed once with water, dried over Na₂SO₄, and the solvent was removed in vacuo. Column chromatography on silica gel (eluent: petroleum ether/ethyl acetate) afforded the desired product.

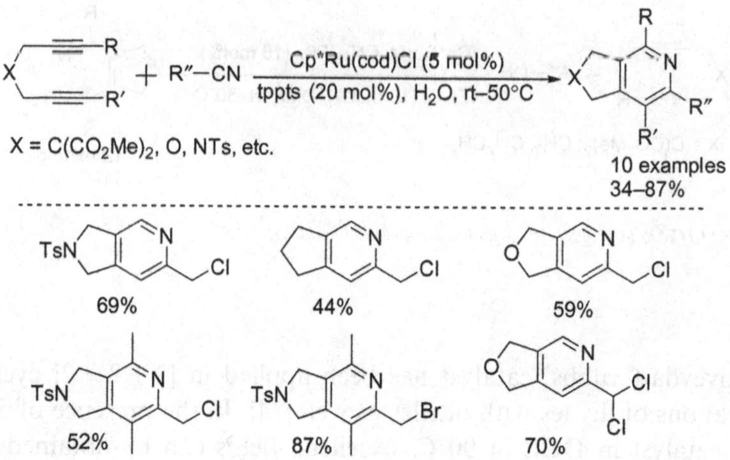

Scheme 3.14 Ru-catalyzed synthesis of pyridines in water.

As the prevailing of pyridine structure in natural chemicals, ruthenium-catalyzed pyridine synthesis has been applied in the total synthesis of natural products as well. In 2011, Witulski and coworkers reported the total synthesis of eudistomin U as a marine natural product [36]. By using functionalized yne-ynamides and methylcyanoformate as the substrates, the desired eudistomin U core moieties can be selectively produced (Scheme 3.15). Detert's group applied this transformation in the total synthesis of perlolyrine and isoperlolyrine in the same year (Scheme 3.16) [37].

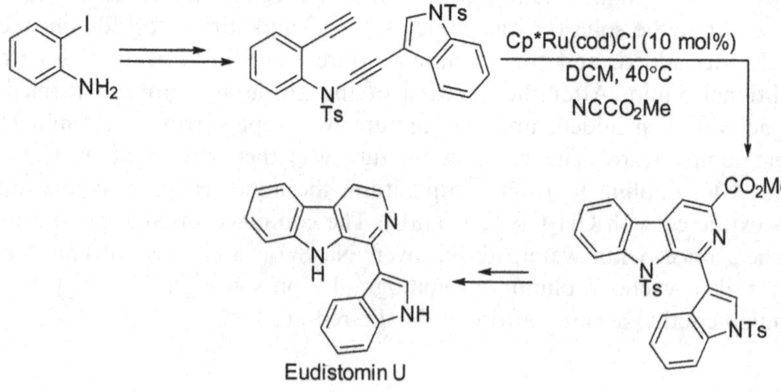

Eudistomin U

Scheme 3.15 Total synthesis of eudistomin U.

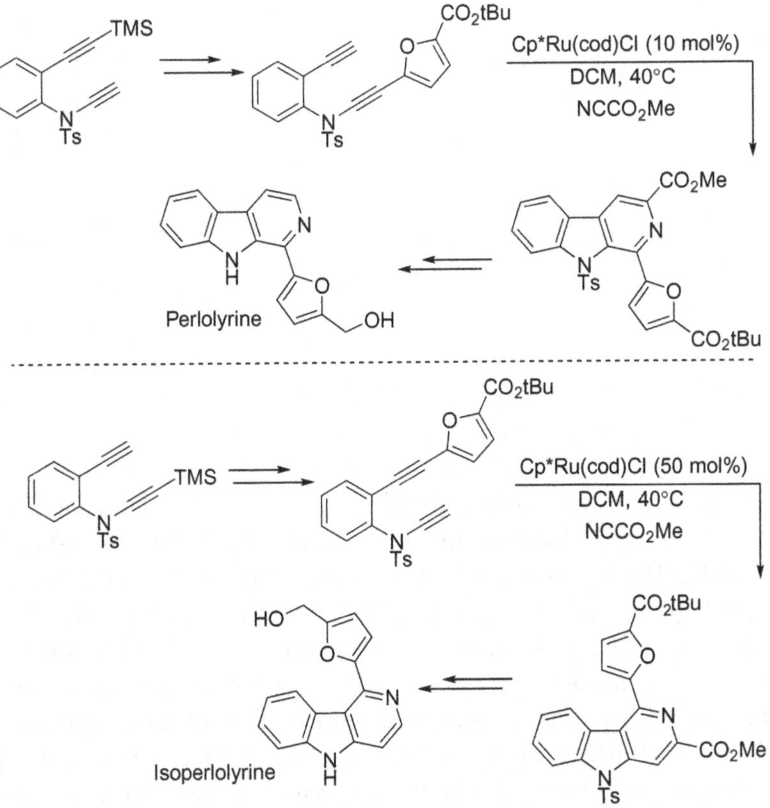

Scheme 3.16 Total synthesis of perlolyrine and isoperlolyrine.

Cyclothiazomycin is 1 of 76 structurally distinct actinomycete thiopeptide antibiotics, can inhibit bacterial protein synthesis, and prevent the growth of gram-positive bacteria [38]. In 2011, Deiters and coworkers reported their work on the synthesis of the pyridine core of cyclothiazomycin based on ruthenium-catalyzed [2 + 2 + 2] cyclotrimerization as the key step (Scheme 3.17) [39]. The electron-deficient nature of the thiazole-bearing nitrile enables ruthenium catalysis under mild reaction conditions with excellent yields. Complete chemo and regioselectivity in the construction of the trisubstituted pyridine core were achieved by applying a temporary silyl tether.

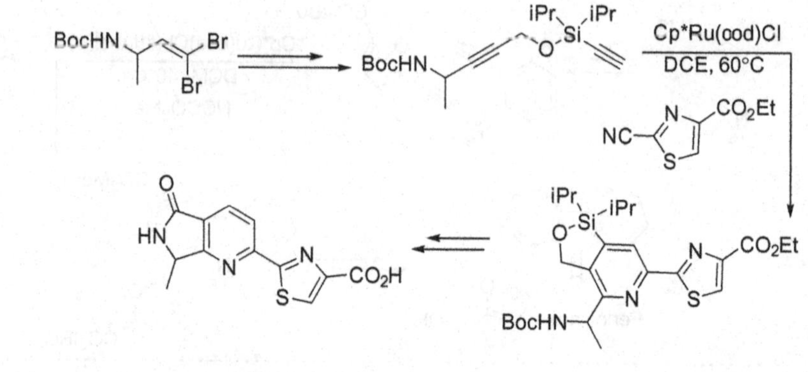

Scheme 3.17 Total synthesis of the pyridine core of cyclothiazomycin.

Rhodium catalysts were applied in the [2 + 2 + 2] cyclotrimerization reactions as well. Tanaka and coworkers found that by using [Rh (cod)$_2$]BF$_4$/BINAP as the catalyst system, highly functionalized pyridines can be produced in good yields from the corresponding alkynes and nitriles under mild reaction conditions (rt-40°C) [40]. Later on, they found by using modified BINAP as the ligand, the reactions can be performed at room temperature with perfluoroalkylacetylenes and perfluoroalkylnitriles as the substrates [41]. Aryl ethynyl ethers [42] and diethyl phosphorocyanidates were found to be suitable substrates as well (Scheme 3.18) [43]. Azatriphenylenes can also be formed by using the proper substrates at room temperature [44].

Scheme 3.18

H$_8$-Binap (6.3 mg, 0.010 mmol) and [Rh(cod)$_2$]BF$_4$ (4.1 mg, 0.010 mmol) were dissolved in CH$_2$Cl$_2$ (2.0 mL), and the mixture was stirred at room temperature for 10 min. H$_2$ was introduced to the resulting solution in a Schlenk tube. The mixture was stirred at room temperature for 1 h, and then the resulting solution was concentrated to dryness.

The residue was dissolved in (CH$_2$Cl)$_2$ (0.8 mL). Compounds alkyne (0.200 mmol) and nitrile (0.220 mmol) were dissolved in (CH$_2$Cl)$_2$ (3.2 mL), and this solution was added to the solution of the above residue. The mixture was stirred at room temperature for 16 h. The resulting mixture was concentrated, and the residue was purified by preparative TLC (hexane/ethyl acetate/triethylamine/methanol, 15:5:10:1) to give the pure product.

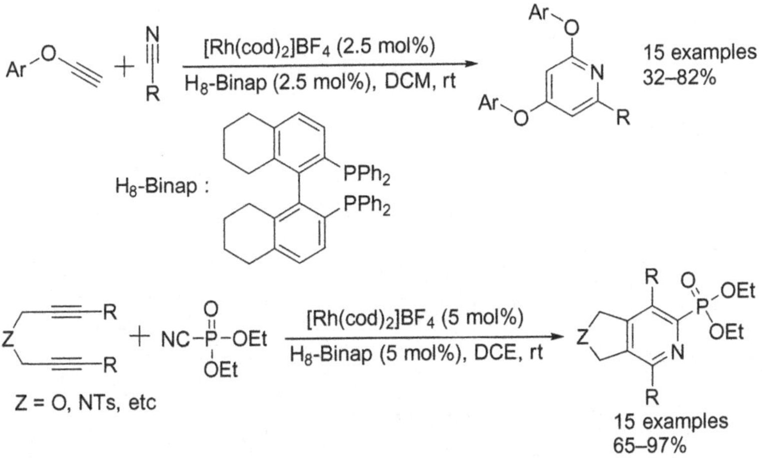

Scheme 3.18 Rh-catalyzed synthesis of pyridines.

Wan and coworkers reported a rhodium-catalyzed [2 + 2 + 2] cyclo-addition of oximes and diynes for the synthesis of pyridines in 2013 [45]. In their mechanistic study, they exclude the dehydration of oxime to generate the corresponding nitrile followed by the cycloaddition of the nitrile and the alkynes to afford the pyridine as the pathway. As only a trace amount of benzonitrile was produced from oxime under the reaction conditions. Additionally, pyridine product was detected in less than 20% yield when benzonitrile was subjected to the reaction with diyne instead of oxime (Scheme 3.19). EtOh was tested as solvent here as well, but not produced. Later on, they found that by using Rh (NBD)$_2$BF$_4$/MeO-Biphep as the catalyst system, the reaction can be performed in EtOH [46].

Scheme 3.19

[Rh(cod)$_2$]BF$_4$ (5.1 mg, 0.0125 mmol) and dppf (8.3 mg, 0.015 mmol) were dissolved in CF$_3$CH$_2$OH (6 mL) in the presence of 4 A molecular sieves, and the mixture was stirred at room temperature for 5 min. Oxime was added and the resulting mixture was stirred at 80°C for 30 min. To this solution was added diyne (0.25 mmol). Then the mixture was stirred at 80°C for 48 h. The 4 A molecular sieves were filtered off and the filtrate was evaporated. The oily residue was purified by column chromatography on silica gel with petroleum ether/ethyl acetate (4:1 to 1:2) as eluent to afford products.

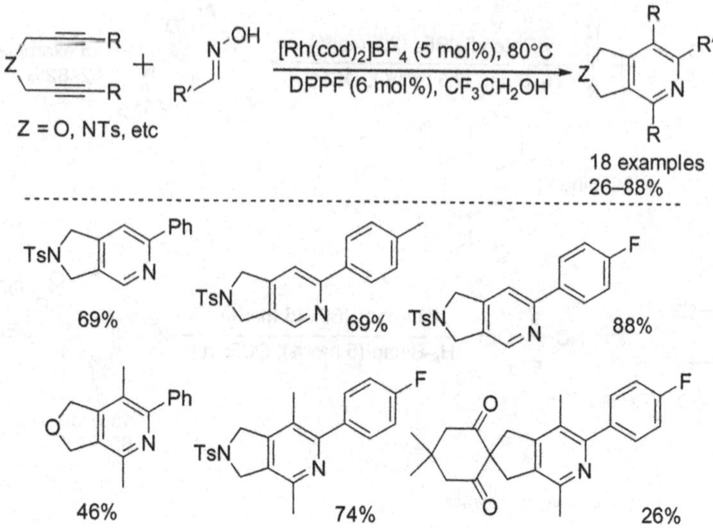

Scheme 3.19 Rh-catalyzed synthesis of pyridines from oxmies.

Iron is cheap, benign, relatively nontoxic, and abundant, and iron catalysis was explored in [2 + 2 + 2] cycloadditions as well. In 2011, Wan and coworkers reported an efficient iron-catalyzed [2 + 2 + 2] cycloaddition of alkynes and cyanamides to the corresponding 2-aminopyridines [47]. The reactions were carried out at room temperature with good to excellent yields (Scheme 3.20). Good regioselectivity and broad substrates scope can be achieved. Interestingly, the presence of both alkynes and nitriles is the key to the generation of active iron species. Additionally, the addition of a small amount of ZnI_2 can improve the reactivity of the catalyst.

Scheme 3.20

FeI_2 (15.6 mg, 0.05 mmol) and dppp (42.4 mg, 0.10 mmol) were weighed in the glove-box and placed in a dried Schlenk tube. Subsequently, distilled THF (2 mL) was added. The resulting mixture was stirred at RT for 30 min to afford an orange/yellow clear solution, at which time Zn dust (6.5 mg, 0.10 mmol) was added. After stirring for an additional 30 min, diyne (0.5 mmol) was added followed by the nitrile (5 mmol), and the reaction was kept stirring for 24 h until the majority of the starting diyne was consumed. The solvent was evaporated and the crude product was directly purified by flash column chromatography on silica gel (eluent: petroleum ether/ethyl acetate) to give the desired pyridine.

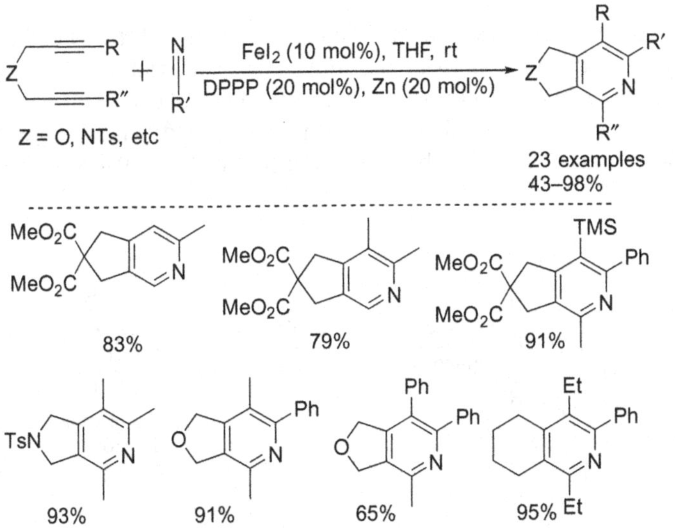

Scheme 3.20 *FeI₂-catalyzed synthesis of pyridines.*

Meanwhile, Louie and coworkers reported a $Fe(OAc)_2$-catalyzed cycloaddition of alkynenitriles and alkynes [48]. Highly substituted pyridines were obtained in good yields with an electron-donating, sterically hindered pyridyl bisimine as the ligand (Scheme 3.21a). Later on, they also showed that diynes and cyanamides can be applied substrates for pyridines synthesis by slightly changing the iron precursor and bisimine ligand (Scheme 3.21b) [49]. Both high yields and regioselectivty can be achieved.

Scheme 3.21a

In a nitrogen-filled glove-box, a solution of alkynenitrile (>1.0 M in DMF) was added to a vial containing 10 mol% $Fe(OAc)_2$ and 13 mol% ligand. Additional DMF was added to make the final concentration of cyanoalkyne 0.4 M (accounting for alkyne volume). The mixture was stirred for 10 min then 1 equiv. of alkyne and 20 mol% of zinc dust were added. The vial was capped and removed from the glove-box then stirred at 85°C for the indicated period of time. The crude mixture was purified via silica gel flash chromatography.

Scheme 3.21b

In a nitrogen filled glove-box, 5 mol% $FeCl_2$, 10 mol% ligand and benzene was added to a vial. The mixture was stirred for 10 to 15 min at which time 1.2 equivalents of cyanamide (2.7 M in benzene) and 10 mol%

Zn dust were added. The vial was then capped with a Teflon lined septum screw cap and removed from the glove-box. The vial was stirred in a 70°C oil bath and a solution of diyne (0.49 M in benzene) was slowly added to the vial over 3 h (unless otherwise noted) via syringe pump. (The final concentration of the diyne after addition was 0.4 M.)

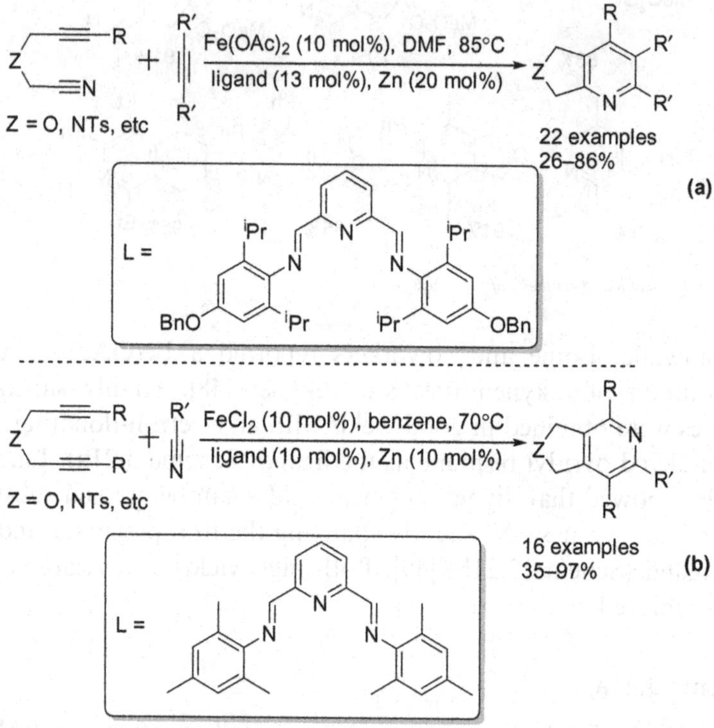

Scheme 3.21 *Fe-catalyzed synthesis of pyridines.*

Recently, Renaud and coworkers reported a new, simple, and air-stable iron(II) complex pre-catalyst catalyzed synthesis of substituted pyridines via [2 + 2 + 2] cycloaddition between diynes and nitrile derivatives [50]. Alkyl-, aryl-, and vinyl nitriles could all be applied; functionalized pyridines can be produced in high yields without any pre-reduction of the catalyst (Scheme 3.22).

Scheme 3.22

In a flamed microwave reactor was added diyne (0.5 mmol.) in the presence of iron catalyst (20 mg, 0.05 mmol, 10 mol%) and tris(2,4,6-trimethoxyphenyl)phosphine (27 mg, 0.05 mmol, 10 mol%). The system was purged with argon. Dry and degassed toluene was added to the solid (840 μL) followed by the degassed nitrile (5 or 40 eq. depending on the boiling point of the nitrile). The reactor vessel was irradiated with microwave (150 W) for 5 min. The solvent was removed under vacuum. The residue was directly purified by flash chromatography on silica gel.

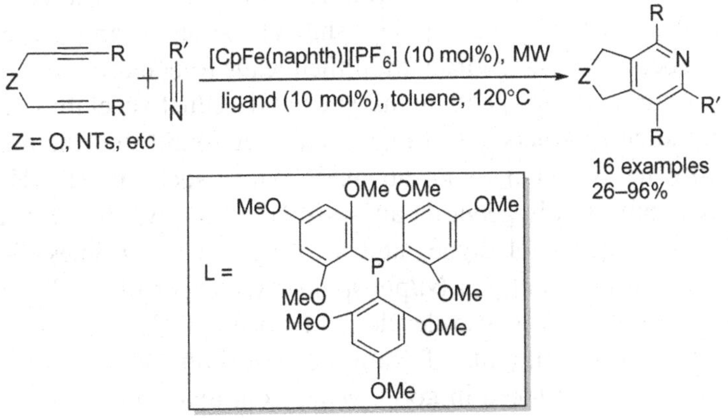

Scheme 3.22 [CpFe(naphth)][PF₆]-catalyzed synthesis of pyridines.

Additionally, several nickel-catalyzed systems for pyridines synthesis have also been developed by Louie and coworkers. In 2005, they reported that with Ni(COD)$_2$ as the catalyst and carbene as the ligand, the cyclization can be carried out at room temperature (Scheme 3.23a) [51]. Both intramolecular and intermolecular reactions were proceeded well and the cycloaddition of an asymmetrical diyne afforded a single pyridine regioisomer. This catalytic system was extended to cyanamides as well [52]. In their systematic studies, they found the in situ formed dimeric [Ni(IPr)RCN]$_2$ from the reaction of Ni(COD)$_2$, IPr, and nitrile are catalytically competent in the formation of pyridines from the cycloaddition of diynes and nitriles. X-ray analysis revealed that these species display simultaneous η^1- and η^2-nitrile binding modes. Kinetic analysis showed the reaction to be first order in [Ni(IPr)RCN]$_2$, zeroth order in added IPr, zeroth order in nitrile, and zeroth order in diyne. Nitrile and ligand exchange experiments were performed and found to be inoperative in the catalytic cycle. These observations suggest a mechanism whereby the catalyst is activated by partial dimer-opening followed by binding of exogenous nitrile and subsequent oxidative hetero-coupling [53]. Reaction profiles demonstrated strong dependence on nitrile, resulting in variable nitrile dependent resting states. The strong coordination and considerable steric bulk of the carbene ligands facilitate selective initial binding of nitrile, thereby forcing a hetero-coupling pathway. In situ IR data suggest that the initial binding of the nitrile resides in a rare, η^1-bound conformation. Following nitrile coordination are a rate-determining hapticity shift of the nitrile and subsequent loss of carbene. Alkyne coordination then leads to heterooxidative coupling, insertion of the pendant alkyne, and reductive elimination to afford pyridine products [54]. In between, they found that the addition of n-BuLi to Ni(acac)$_2$ and an NHC salt (such as IPr·HCl or SIPr·HCl) can rapidly generate an Ni(0)/NHC catalyst which is active for the cycloaddition of diynes and nitriles to give pyridines (Scheme 3.23b) [55]. More recently, a Ni/phosphine system for the cyclization of alkynes and nitriles was also developed (Scheme 3.23c). By the combination of catalytic amounts of Xantphos and Ni(COD)$_2$, a variety of pyridines can be produced in good yields from unactivated nitriles and diynes under mild reaction conditions [56].

Scheme 3.23a

In a nitrogen-filled glove-box, a solution of Ni(COD)$_2$ and SIPr in toluene were allowed to equilibrate for at least 6 h. A solution of diyne and nitrile in toluene were placed into an oven-dried vial equipped with a stir bar. To this stirred solution was added the solution of Ni(COD)$_2$ and SIPr. The reaction mixture was stirred for 2 h at ambient temperature, taken out of the glove-box, and concentrated under diminished pressure. The remaining residue was purified with flash column chromatography to afford the pyridine. This procedure was used for the cycloaddition reactions unless otherwise noted.

Scheme 3.23b

To a stirring suspension of Ni(acac)$_2$ and (S)IPr·HCl in hexanes was added n-BuLi (2.5 M in hexanes) dropwise at room temperature. The resulting suspension was stirred for an additional 5 min at which time a solution of diyne (0.2 M in hexanes or toluene) was added followed by the nitrile. The reaction mixture was stirred at room temperature until no starting diyne was detected by TLC analysis (30 min). The reaction mixture was then quenched by the addition of five drops of MeOH and concentrated in vacuo. The residue was purified by flash column chromatography to yield the desired pyridine.

Scheme 3.23c
In a nitrogen-filled glove-box, diyne (1 equiv., 0.1 m) and nitrile (1.5 equiv.) were added to an oven-dried screw-cap vial equipped with a magnetic stir bar. In a separate vial [Ni(cod)$_2$] and Xantphos were weighed out (in 1:1 molar ratio) and dissolved in toluene. The catalyst (3 mol%) solution was added to the reaction mixture. The vial was sealed and brought out of the glove-box. The reaction was stirred at RT for 3 h. The resulting reaction mixture was concentrated and purified by flash column chromatography on silica gel using first 15%, 30%, and finally 50% EtOAc/hexanes.

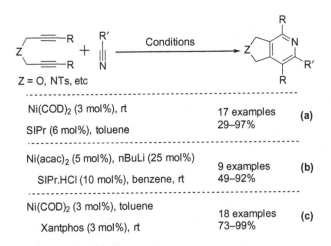

Scheme 3.23 Ni-catalyzed synthesis of pyridines.

Takeuchi and coworkers reported an iridium-catalyzed cycloaddition of α,ω-diynes and nitriles to give pyridines in 2012 [57]. With [Ir (cod)Cl]$_2$/DPPF or BINAP as the catalytic system, pyridines were formed effectively (Scheme 3.24). A wide range of nitriles (aliphatic and aromatic nitriles) can be applied and reacted smoothly with α,ω-diynes to give the pyridines. In the case of unsymmetrical diyne bearing two different internal alkyne moieties, high regioselectivity can be achieved which can be explained by the different reactivities of the α-position in iridacyclopentadiene. Terpyridine and quinquepyridine were prepared as well. [Ir(cod)Cl]$_2$/chiral diphosphine catalyst can be applied to enantioselective synthesis. Kinetic resolution of the racemic secondary benzyl nitrile catalyzed by [Ir(cod)Cl]$_2$/SEGPHOS gave a central carbon chiral pyridine in 80%. Density functional calculations were also performed for understanding the reaction mechanism.

Scheme 3.24

A flask was charged with [Ir(cod)Cl]$_2$ (7.2 mg, 0.01 mmol) and DPPF (11.3 mg, 0.02 mmol). The flask was evacuated and filled with argon. To the flask were added benzene (5 mL) and benzonitrile (325 mg, 3.2 mmol). Diyne (1.0 mmol) was added to the reaction mixture. The mixture was stirred under reflux for 3 h. The progress of the reaction was monitored by GLC. After the reaction was complete, the solvent was evaporated in vacuo. Column chromatography of the residue gave pure product (n-hexane/AcOEt = 70/30).

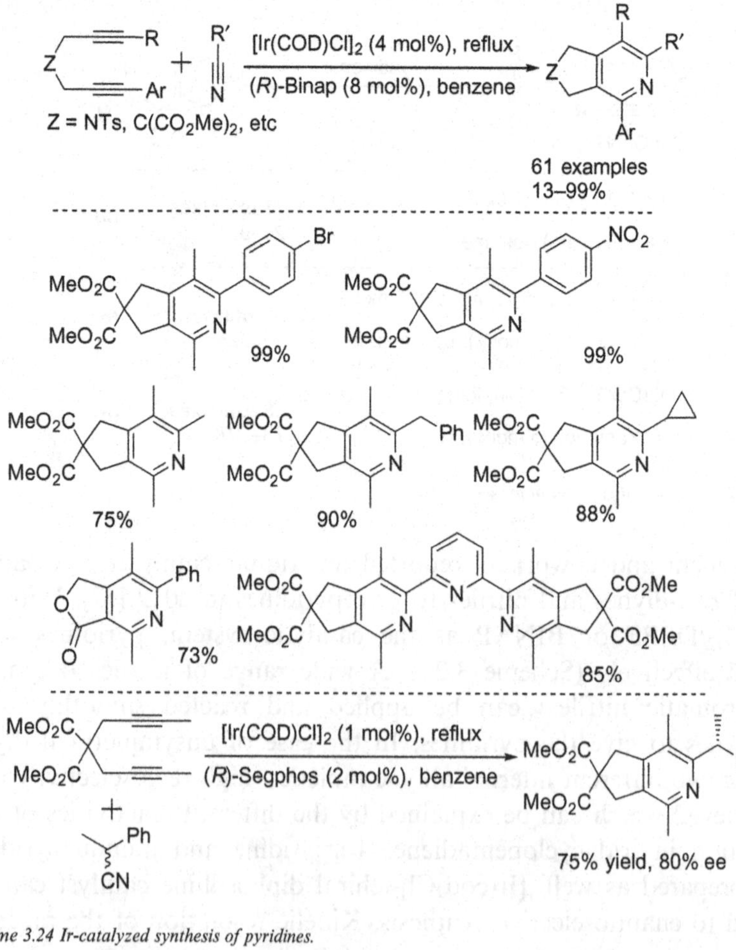

Scheme 3.24 Ir-catalyzed synthesis of pyridines.

Obora and Satoh showed that low-valent niobium can be applied as catalyst in the cycloaddition of alkynes and nitriles as well [58]. By the combination of $NbCl_5$, Zn, and an alkoxysilane, pyridine derivatives were produced in good yields from the intermolecular cycloaddition of alkynes and nitriles via a niobacyclopentadiene intermediate (Scheme 3.25). Zn acts as a reducing agent here, the active low-valent Nb species might also be stabilized by the alkoxysilane, and generated chloro (methoxy)diphenylsilane.

Scheme 3.25

A mixture of *tert*-butylacetylene (82 mg, 1 mmol), benzonitrile (309 mg, 3 mmol), $NbCl_5$ (54 mg, 0.2 mmol), Zn (78 mg, 1.2 mmol), $Ph_2Si(OMe)_2$ (146 mg, 0.6 mmol), and toluene (2 mL) was stirred for 16 h at 80°C under Ar. The yields of the products were estimated from the peak areas on the basis of the internal standard technique using GC. The products were isolated by silica gel column chromatography (*n*hexane: EtOAc = 100:0 to 50:1 as eluent).

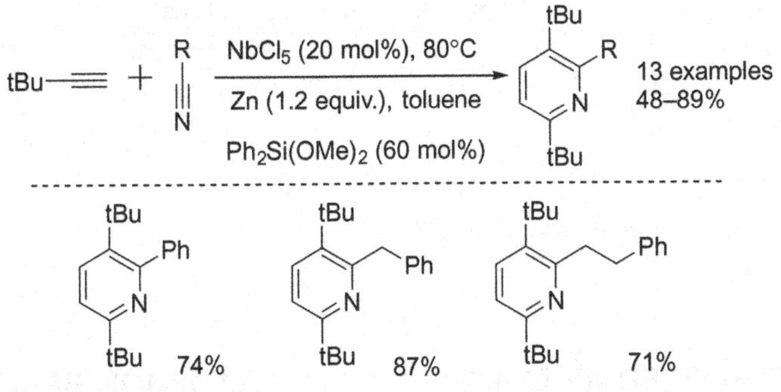

Scheme 3.25 Nb-catalyzed synthesis of pyridines.

Additionally, Pd-catalyzed cross-coupling reactions were explored in pyridines synthesis also. In 2007, a multicomponent sequential process for the synthesis of trisubstituted pyridines was reported [59]. The reaction involves the formation of an enamine by Pd-catalyzed amination of an alkenyl bromide, formation of a 2-aza-1,3-butadiene by Pd-catalyzed cross-coupling of a trimethylsilylimine with an alkenyl bromide, and Lewis acid ($Yb(OTf)_3$ (20 mol%)) catalyzed cycloaddition between the enamine and the azadiene. Moderate yields can be achieved (Scheme 3.26). A zinc-mediated tandem reaction of nitriles with propargyl bromides to give 3-alkynylpyridines was developed by Fan and coworkers as well [60]. The reaction works at room temperature in THF in the presence of two equivalents of zinc powder.

Scheme 3.26

A carousel reaction tube under nitrogen atmosphere was charged with Pd₂(dba)₃ (0.01 mmol, 1 mol%), DavePhos (0.04 mmol, 4 mol%), NaOtBu (1.1 mmol), and toluene (1 mL). After 1 min of stirring, morpholine (0.5 mmol), the bromoalkene (1 mmol), the silylimine (0.6 mmol), and Yb(OTf)₃ (0.1 mmol, 20 mol%) were added sequentially while stirring, separated by 1 min intervals. Finally an additional 1 mL of toluene was added. The mixture was heated at 90°C for 14 h. Once the reaction has been completed, the mixture is allowed to reach rt, diluted with EtOAc (15 mL), and washed with saturated NaHCO₃ solution. The aqueous phase is extracted with EtOAc (2 × 5 mL) and the combined organic layers are dried over Na₂SO₄ and concentrated under reduced pressure. The pyridines are purified by flash chromatography.

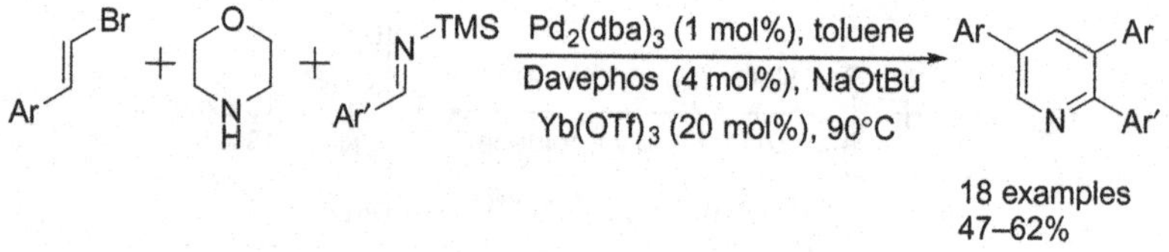

Scheme 3.26 Pd-catalyzed multicomponent synthesis of pyridines.

3.2 SYNTHESIZED BY [2 + 2 + 1 + 1] CYCLIZATION REACTIONS

In 2015, Deng and coworkers reported a ruthenium-catalyzed one-pot procedure for the synthesis of 2,4-diarylsubstituted pyridines [61]. They reported that using acetophenones, ammonium acetate, and DMF as the substrates and under an oxygen atmosphere, good yields of the corresponding pyridines can be achieved (Scheme 3.27). Notably, the carbon atom situated at C6 of the pyridine ring comes from the methyl group of DMF and the nitrogen atom comes from ammonium acetate. Soon later, Yuan and coworkers reported a NH₄I-promoted cyclization of ketones with DMSO and NH₄OAc to give pyridines at 130°C [62]. In their deuterium-labeling experiments, they found the C4 or C6 of the target product pyridine rings resulted from DMSO.

Scheme 3.27

A 20 mL reaction vessel was charged with $RuCl_3 \cdot 3H_2O$ (2.6 mg, 0.01 mmol), acetophenone (48 μL, 0.4 mmol), NH_4OAc (46.2 mg, 0.6 mmol). The sealed reaction vessel was purged with oxygen three times. DMF (0.5 mL) was added to the sealed reaction vessel by syringe. The resulting solution was stirred at 120°C for 24 h. After cooling to room temperature, the volatiles were removed under vacuum and the residue was purified by column chromatography (silica gel, petroleum ether/dichloromethane = 3:1) to give the pure product.

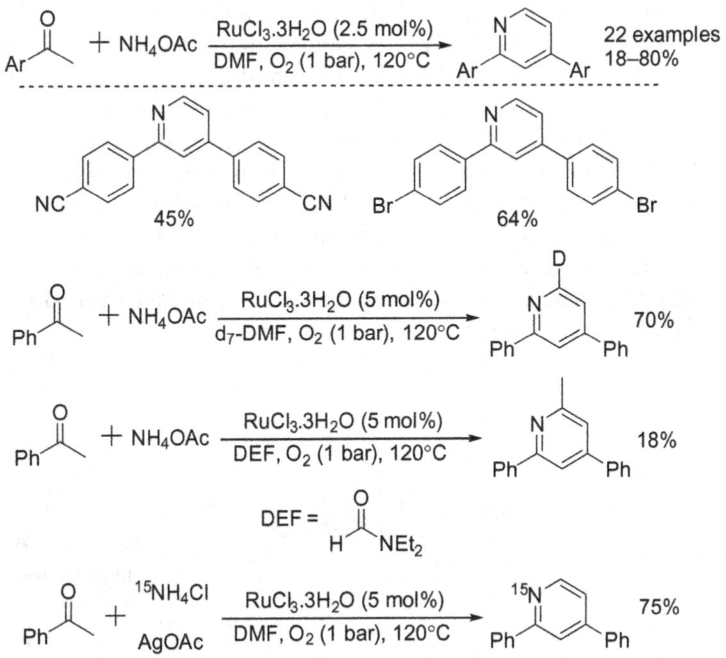

Scheme 3.27 Ru-catalyzed four components synthesis of pyridines.

In the same year, Jiao and coworkers developed a Cu-catalyzed synthesis of pyridines from acetaldehydes and nitrogen source [63]. 3,5-Diarylpyridines were produced in good cascade Chichibabin-type cyclization, C-(sp³)-C(sp³) cleavage, and aerobic oxidation (Scheme 3.28). Azide and ceric ammonium nitrate (CAN) were found as efficient nitrogen donors and with O_2 as the oxidant.

Scheme 3.28

To a reaction tube charged with copper trifluoroacetate hydrate (10.9 mg, 0.0375 mmol, 15 mol%) and NHPI (8.2 mg, 0.05 mmol, 0.2 equiv.) was added a solution of acetaldehyde (1 mmol), TMSN$_3$ (33 uL, 0.25 mmol, 1 equiv.), acetic acid (14 uL, 0.25 mmol, 1 equiv.), and H$_2$O (135 μL, 7.5 mmol, 30 equiv.) in DMF (3 mL) under O$_2$ (1 atm). The reaction mixture was then stirred at 80°C for 24 h. After cooling to room temperature, the mixture was diluted with ethyl acetate, washed with saturated sodium bicarbonate, water and brine, dried over anhydrous sodium sulfate, and concentrated in vacuo to give dark residue, which was purified by flash chromatography (using petroleum ether and ethyl acetate as the effluent) on silica gel to afford the 2-ketopyridine product.

To a reaction tube charged with cerium ammonium nitrate (68 mg, 0.125 mmol, 0.25 mmol of NH$_4^+$, 1 equiv.) and copper trifluoroacetate hydrate (14.5 mg, 0.05 mmol, 20 mol%) was added a solution of acetaldehyde (1 mmol) and H$_2$O (135 μL, 7.5 mmol, 30 equiv.) in DMF (3 mL) under argon (1 atm). The reaction mixture was then stirred at 80°C for 12 h. After cooling to room temperature, the mixture was diluted with ethyl acetate, washed with saturated sodium bicarbonate, water and brine, dried over anhydrous sodium sulfate, and concentrated in vacuo to give dark residue, which was purified by flash chromatography (using petroleum ether and ethyl acetate as the effluent) on silica gel to afford the 5-diarylpyridine product.

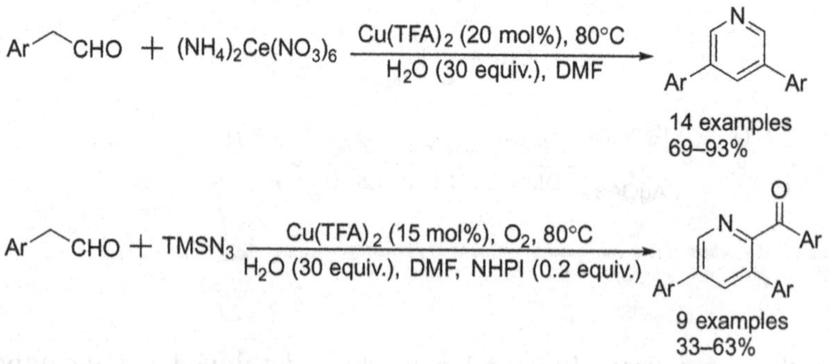

Ar⁀CHO + (NH$_4$)$_2$Ce(NO$_3$)$_6$ $\xrightarrow{\text{Cu(TFA)}_2 \text{ (20 mol%), 80°C}}_{\text{H}_2\text{O (30 equiv.), DMF}}$

14 examples
69–93%

Ar⁀CHO + TMSN$_3$ $\xrightarrow{\text{Cu(TFA)}_2 \text{ (15 mol%), O}_2, \text{ 80°C}}_{\text{H}_2\text{O (30 equiv.), DMF, NHPI (0.2 equiv.)}}$

9 examples
33–63%

Scheme 3.28 Cu-catalyzed four components synthesis of pyridines.

3.3 SYNTHESIZED BY [3 + 3] CYCLIZATION REACTIONS

A tandem one-pot method for the construction of highly substituted pyridine derivatives from nitriles, Reformatsky reagents, and 1,3-enynes was developed by Lee and coworkers in 2011 [64]. The reaction sequential consisted of the reaction of nitrile with a Reformatsky reagent and then selective addition of the Blaise reaction intermediate to 1,3-enyne, followed by isomerization, cyclization, and an aromatization cascade to give the final pyridines in moderate to excellent yields.

Scheme 3.29

Method A: A 10 mL Schlenk tube equipped with a stirrer bar was charged with oxime (0.2 mmol), pyrrolidinium perchlorate (6.9 mg, 0.04 mmol, 20 mol%), and CuI (7.6 mg, 0.04 mmol, 20 mol%). The Schlenk tube was quickly evacuated and refilled with N_2 three times, followed by the addition of unsaturated aldehyde (0.3 mmol, 1.5 equiv.) and DMSO (1 mL). The Schlenk tube was sealed with a Teflon screwcap and then the reaction mixture was stirred at 60°C for 16 h. Upon cooling to room temperature, the reaction mixture was diluted with 5 mL of ethyl acetate and filtered through a pad of silica gel with additional ethyl acetate (15 mL) as the eluent. The filtrate was washed with water (10 mL), dried over Na_2SO_4, and concentrated under reduced pressure. The residue was purified by flash chromatography on silica gel to afford the pyridine derivative.

Method B: A 10 mL Schlenk tube equipped with a stirrer bar was charged with oxime (0.2 mmol) and CuI (7.6 mg, 0.04 mmol, 20 mol%). The Schlenk tube was quickly evacuated and refilled with N_2 three times, followed by the addition of unsaturated aldehyde (0.3 mmol, 1.5 equiv.), diisopropylamine (56 μL, 0.4 mmol, 2 equiv.), and DMSO (1 mL). The Schlenk tube was sealed with a Teflon screwcap and then the reaction mixture was stirred at 60°C for 16 h. Upon cooling to room temperature, the reaction mixture was diluted with 5 mL of ethyl acetate and filtered through a pad of silica gel with additional ethyl acetate (15 mL) as the eluent. The filtrate was washed with water (10 mL), dried over Na_2SO_4, and concentrated under reduced pressure. The residue was purified by flash chromatography on silica gel to afford the pyridine derivative.

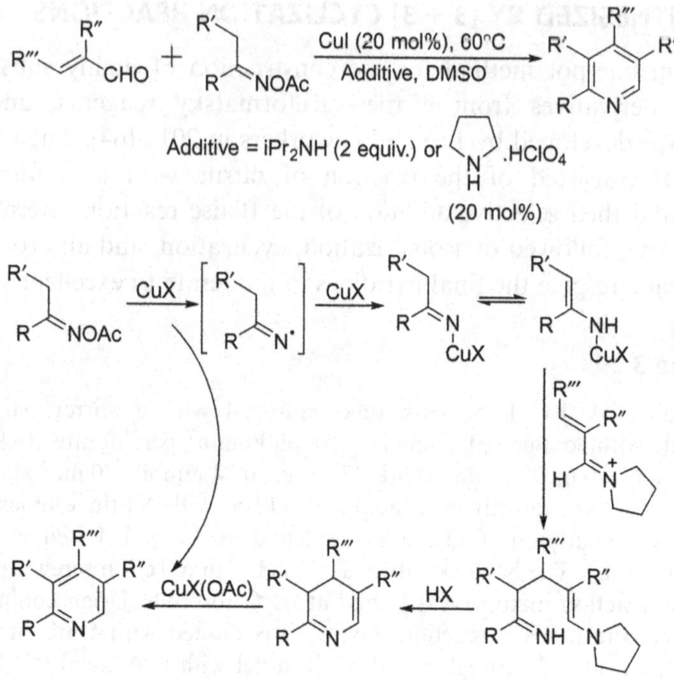

Scheme 3.29 Cu-catalyzed synthesis of pyridines from oximes and enals.

In 2013, Yoshikai and Wei developed a copper-catalyzed pyridines synthesis from oximes and enals [65]. Under redox-neutral reaction conditions, with O-acetyl ketoximes and α,β-unsaturated aldehydes as the substrates and using copper(I) salt and a secondary ammonium salt (or amine) as the catalyst system, a variety of substituted pyridines were prepared with a broad range of functional groups tolerance (Scheme 3.29). By merging iminium catalysis and copper catalyst, under the redox activity of the copper catalyst, the reaction started to reduce the oxime N—O bond to generate a nucleophilic copper(II) enamide and later oxidize a dihydropyridine intermediates to the final products.

Arcadi and coworkers developed a one-pot procedure for the synthesis of functionalized pyridines from carbonyl compounds and propargylamine [66]. Followed by sequential amination/annulation/aromatization reactions, good yields of the desired pyridines can be isolated (Scheme 3.30). Steroidal derivatives with fused pyridine rings and polyannelated pyridines can be selectively produced by this procedure as well. By using gold as the catalyst, linear polycyclic pyridines were regioselectively produced and angular polycyclic pyridines can be obtained from the corresponding α,β-unsaturated derivatives.

Scheme 3.30

To a 50 mL stainless steel autoclave charged with a solution of the ketone (1.26 mmol) in absolute ethanol (5 mL) were added propargylamine (2.52 mmol) and the catalyst (0.03 mmol). The resulting mixture was heated at reflux under stirring or at 100 or 140°C. The reaction was monitored by TLC and GC-MS. After cooling, the mixture was filtered to remove the catalyst, and the solvent was concentrated under reduced pressure. The residue was purified by flash chromatography (silica gel, n-hexanes-ethyl acetate mixtures) to give pyridine. Alternatively, the residue was dissolved in ethyl acetate and extracted three times with a 6 M solution of aqueous hydrochloric acid. The combined aqueous extracts were washed with ethyl acetate, solid NaOH was added until pH = 8, and the mixture was extracted three times with ethyl acetate. The combined organic extracts were dried (Na_2SO_4) and evaporated under reduced pressure. The desired product was distilled under vacuum.

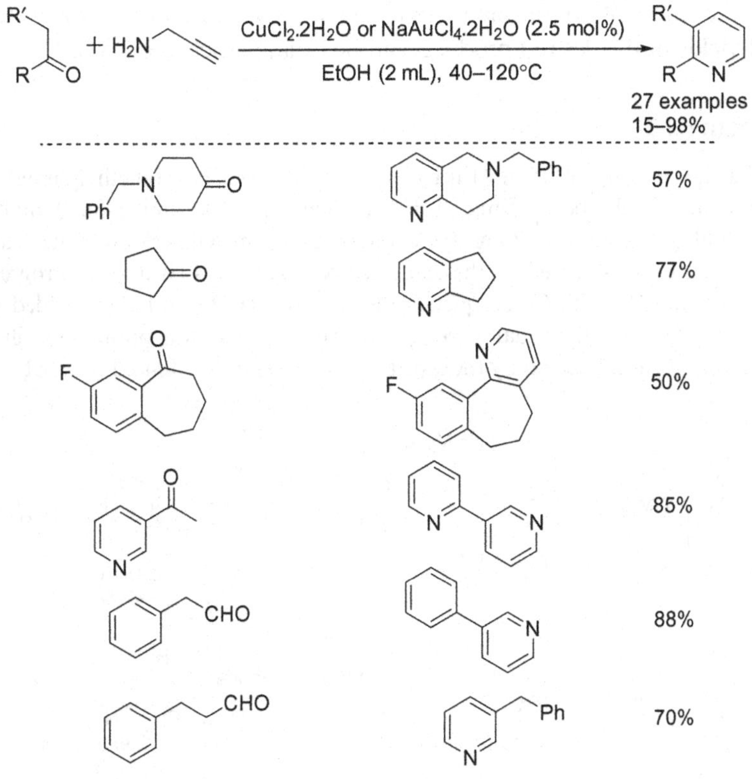

Scheme 3.30 Au/Cu-catalyzed synthesis of pyridines from carbonyl compounds and propargylamine.

The using of β-ketoester, ammonia, and alkynone as substrates, with Brønsted acid (AcOH) or Lewis acid (ZnBr$_2$) or Amberlyst 15 ion exchange resin as the promotor, to synthesis 2,3,6-trisubstituted or 2,3,4,6-tetrasubstituted pyridines was also reported [67]. Good yields and total regiocontrol can be obtained.

A ytterbium triflate (Yb(OTf)$_3$) promoted production of pyridine derivatives by the cyclization of N-silylenamine and in situ generated 2-methylene-1,3-cyclohexanedione and 2-methylenecyclohexanone was developed in 2006 [68]. Notably, 69% of the desired product can still be produced in the absence of ytterbium triflate. For the reaction mechanism, the reaction started with nucleophilic attack of the β-carbon of the enamine on the β-position of the α,β-unsaturated carbonyl group, which was activated by the ytterbium salt, and followed by an in situ keto-enol isomerization to form the corresponding diketone intermediate. Then intramolecular attack of nitrogen atom in the intermediate on the activated carbonyl group occurs to give the cyclization product, which will give the final product after the elimination of silanol and subsequent oxidation. Moderate to good yields of the desired products can be isolated (Scheme 3.31).

Scheme 3.31

To a 1,4-dioxane solution (1 mL) of [(2,6-dioxocyclohexyl)methyl]dimethylammonium chloride (125 mg, 0.60 mmol) and N-silylenamine (0.50 mmol) was added Yb(OTf)$_3$ (6.2 mg, 0.010 mmol) at room temperature. The reaction mixture was stirred at the same temperature for 40 h and its progress was monitored by TLC. To quench the reaction, H$_2$O (2 mL) was added to the mixture. After the usual work-up, the crude product was purified by silica gel column chromatography (hexane/EtOAc = 1:1) to the desired product.

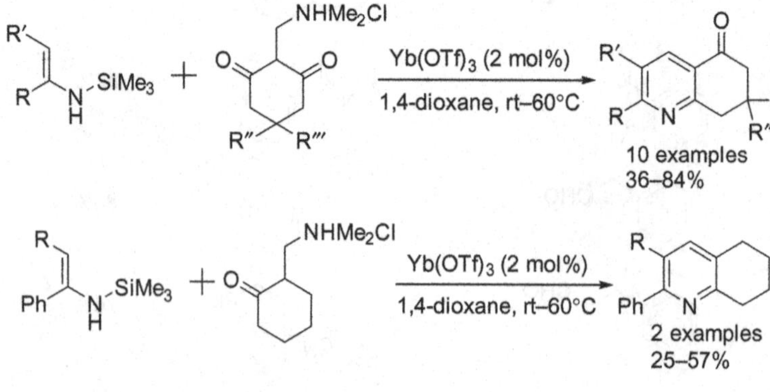

Scheme 3.31 Yb-catalyzed synthesis of 1,3-disubstituted 7,8-dihydroquinolin-5-one.

Peters and coworkers reported a regioselective palladium-catalyzed synthesis of 2,3,6-trisubstituted pyridines with isoxazolinones and vinyl ketones as the substrates [69]. This protocol involves two palladium-catalyzed catalytic cycles. The first cycle is a Pd(II)-catalyzed *C*-regiose-lective 1,4-addition of isoxazolinones to vinylketones. Then followed by a Pd(0)-catalyzed transformation, which is assumed to proceed via vinylnitrene-palladium intermediates, to give the final pyridine deriva-tives. Both hydrogen and air as reduction and oxidation reagents are nec-essary for the pyridine formation. Here, the ratios of hydrogen and air are above the upper explosive limit thus avoiding a safety issue. Moderate to good yields of the desired pyridines were formed (Scheme 3.32).

Scheme 3.32

General Procedure cycle 1. To a solution of the corresponding isoxazol-5 (4*H*)-one (1 equiv., recrystallized from EtOH prior to use) in CH_2Cl_2 (0.6 M) was added palladium(II) acetate (1−10 mol%) and subsequently the corresponding vinyl ketone (4 equiv.). After 1 h of stirring at room temperature, the solvent was removed and the crude product was sub-jected to column chromatography (silica, petrol ether/ethyl acetate: 4/1).

General Procedure cycle 2. To a solution of the 1,4-adduct (1 equiv.) in 1,4-dioxane was added tetrakis(triphenylphosphine)palladium(0) (0.5−5 mol %) as stock solution, so that a substrate concentration of 0.1 M was achieved. Hydrogen (over a balloon) was bubbled through the solution for 10 min. Thereafter the solution was set under a hydrogen atmosphere and it was stirred for 24 h at room temperature. The solvent was evaporated and the crude product was subjected to column chromatography (silica, petrol ether/ethyl acetate: 4/1) to give the respective pyridine.

In 2009, Chiba's group reported a manganese-mediated synthesis of pyridines from cyclopropanols with vinyl azides [70]. In the presence of 1.7 equivalents of $Mn(AcAc)_3$, a variety of pyridine derivatives have been prepared in good yields at room temperature. In their further studies, they realized this transformation in a catalytic manner [71]. In their mechanistic studies, they found the reactions were initiated by a radical addition of β-carbonyl radicals, generated by the one-electron oxidation of cyclopropanols with Mn(III), to vinyl azides to give imi-nyl radicals, which cyclized with the intramolecular carbonyl groups. Additionally, the application of this newly developed methodology in the synthesis of quaternary indole alkaloid, melinonine-E, was accom-plished as well (Scheme 3.33).

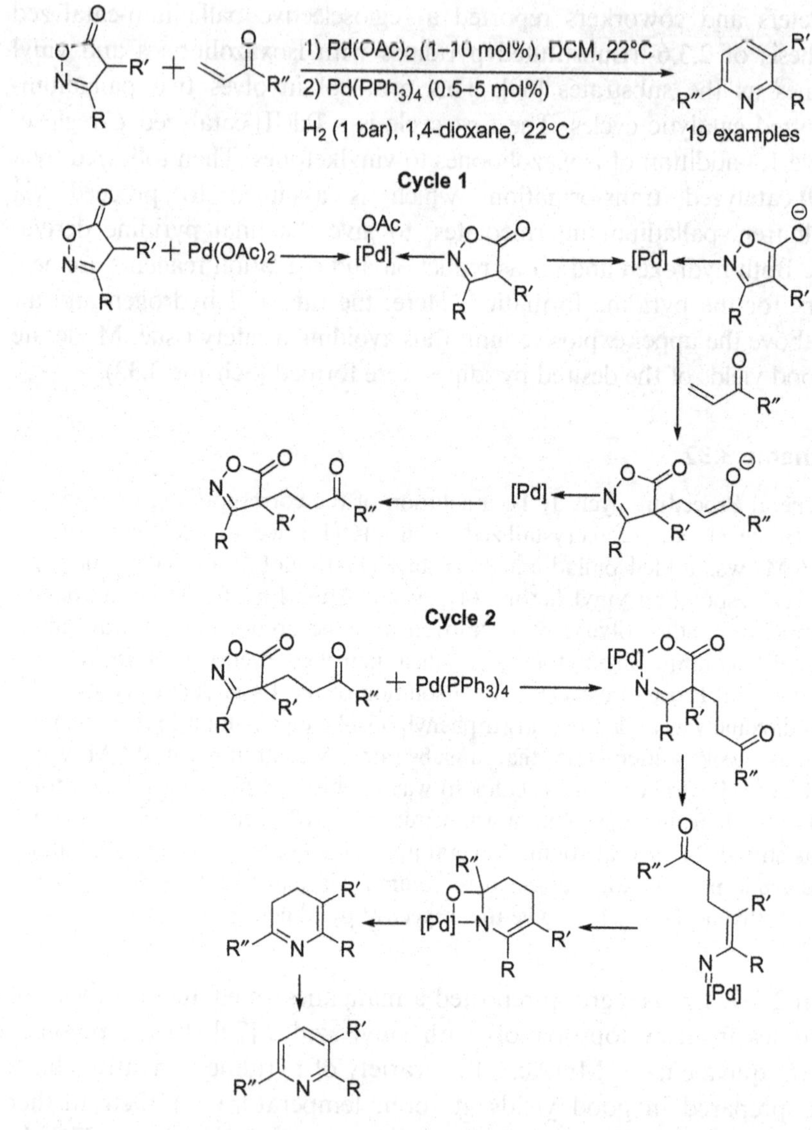

Scheme 3.32 Pd-catalyzed synthesis of pyridines from isoxazolinones and vinylketones.

Scheme 3.33

Stoichiometric reaction. To a solution of α-azidostyrene (43.5 mg, 0.30 mmol) and 1-phenylcyclopropanol (60.6 mg, 0.45 mmol) in MeOH (3.0 mL) was added Mn(acac)$_3$ (179.8 mg, 0.51 mmol) at room temperature under nitrogen atmosphere. After 5 min, AcOH (34 μL, 0.60 mmol) was added and the reaction mixture was stirred for 1 h at the same temperature. The reaction mixture was quenched with pH 9 ammonium buffer and then extracted twice with ethyl acetate. The combined organic extracts were washed with brine, dried over MgSO$_4$, and concentrated. Purification of the crude product by flash column chromatography (silica gel; hexane : ethyl acetate = 98:2) afforded pure product.

Catalytic reaction. To a solution of α-azidostyrene (43.6 mg, 0.30 mmol) and 1-phenylcyclopropanol (48.4 mg, 0.36 mmol) in MeOH (3.0 mL) was added Mn(acac)$_3$ (10.6 mg, 0.03 mmol) at room temperature under nitrogen atmosphere. After 5 min, HCl (0.20 mL, 0.60 mmol, 3.0 M in MeOH) was added and the nitrogen balloon was then replaced by oxygen balloon. The reaction mixture was heated at 40°C for 1 h and quenched with pH 9 ammonium buffer, and then extracted twice with ethyl acetate. The combined organic extracts were washed with brine, dried over MgSO$_4$, and concentrated. Purification of the crude product by flash column chromatography (silica gel; hexane : ethyl acetate = 98:2) afforded the pure product.

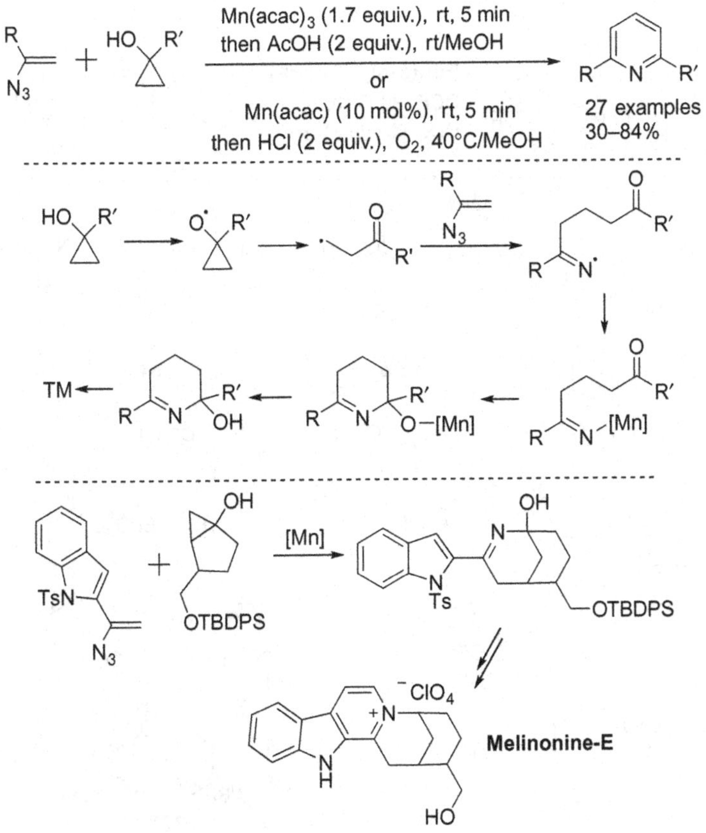

Scheme 3.33 Mn-catalyzed synthesis of pyridines from cyclopropanols with vinyl azides.

Park and coworkers developed a rhodium-catalyzed synthesis of pyridines from 2*H*-azirines and carbenoids [72]. In the presence of catalytic amount of [Rh$_2$(esp)$_2$], highly substituted pyridines were produced in good to excellent yields (Scheme 3.34). This procedure was applied by Liang and coworkers in the synthesis of CF$_3$-containing pyridines recently [73].

Scheme 3.34

An oven dried Schlenk tube charged with [Rh$_2$(esp)$_2$] (0.02 equiv.) was purged with nitrogen, and a solution of azirine (0.3 mmol, 1 equiv.) in DCE (1 mL) was added. A solution of the freshly prepared diazo compound (1.6 equiv.) in DCE (1 mL) was added dropwise to the suspension under nitrogen. The reaction mixture was then heated to 90°C for 3–20 h. The solution was cooled to RT and treated with DDQ (1 equiv.). The suspension was stirred at RT for 15 min and filtered through a plug of silica gel. The filtrate was concentrated in vacuo, and the residue was purified by flash chromatography using hexanes/ethyl acetate (4:1) as the eluent to give the desired product.

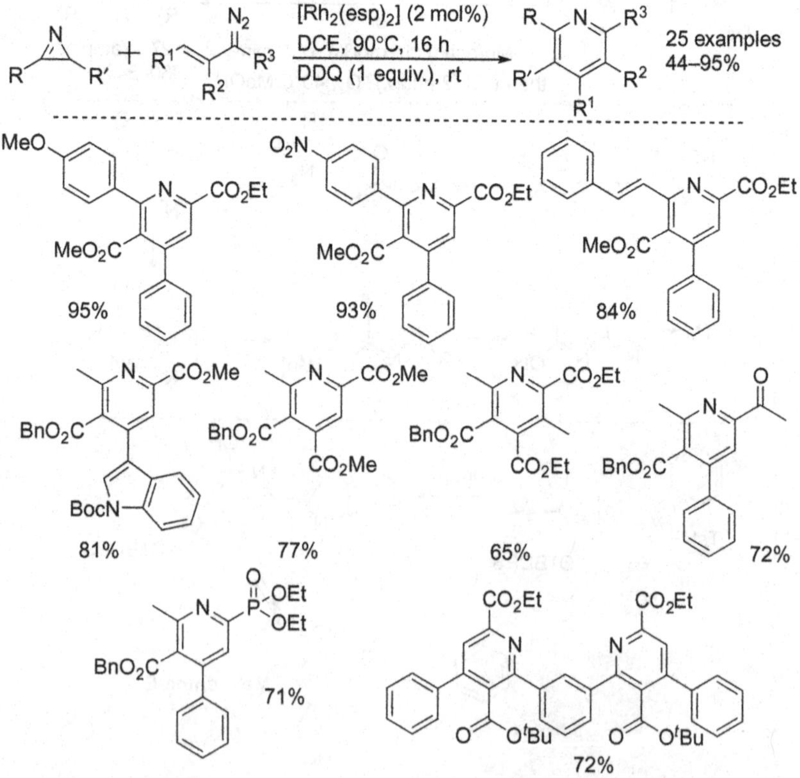

Scheme 3.34 Rh-catalyzed synthesis of pyridines from 2H-azirines and carbenoids.

Davies and coworkers discovered that pyridines can be produced by rhodium-catalyzed cyclization of carbenoids and isoxazoles as well [74]. Highly functionalized pyridines and 1,4-dihydropyridines could be produced by this procedure in good yields (Scheme 3.35). For the reaction mechanism, the reaction proceeds through an initial carbenoid induced ring expansion of isoxazoles and then followed by a rearrangement/tautomerization/oxidation sequence.

Scheme 3.35

To a flame dried round-bottomed flask charged with a stir bar and under argon was added 3,5-dimethylisoxazole (0.049 g, 0.50 mmole), $Rh_2(OAc)_4$ (1.1 mg, 0.0025 mmol, 0.005 equiv.) and toluene (1.5 mL). A reflux condenser was attached to the flask and the solution was heated in an oil bath to 60°C. A solution of (*E*)-methyl 2-diazo-4-phenylbut-3-enoate (0.202 g, 1.0 mmol) in toluene (1.5 mL) was then added dropwise into the solution by syringe pump over 30 min. The solution was then immediately transferred to a heating mantle and heated at reflux for 4 h. The solution was then allowed to cool to room temperature and DDQ (0.114 g, 0.5 mmol) was added. The solution was stirred for 30 min at room temperature, and then diluted with a solution of diethyl ether containing 1% triethylamine (v/v, 10 mL). The solution was vacuum filtered through a plug of silica gel (1.5" x 1.5") using the 1% triethylamine solution in diethyl ether to wash through (200 mL). The filtrate was concentrated in vacuo and purified by flash chromatography to give the pure product.

3.4 SYNTHESIZED BY [3 + 2 + 1] CYCLIZATION REACTIONS

Among all the pyridine derivatives, 2-amino-3,5-dicyano-6-sulfanyl pyridines are even more interesting because of their potential therapeutic applications. In synthetic chemistry, the cyclocondensation of aldehydes, malononitrile, and thiols is the most straightforward pathway. In general, this transformation can be realized under basic conditions. The bases reported included Et_3N, DABCO, piperidine, morpholine, thiomorpholine, pyrrolidine, *N,N*-DIPEA, pyridine, 2,4,6-collidine, DMAP, aniline, *N*-methylaniline, *N,N*-dimethylaniline and *N,N*-diethylaniline. Reactions under neutral conditions have been explored as well. In these cases, CuI nanoparticles [75], $ZnCl_2$ [76], and nanocrystalline magnesium oxide [77] have been applied as the catalysts.

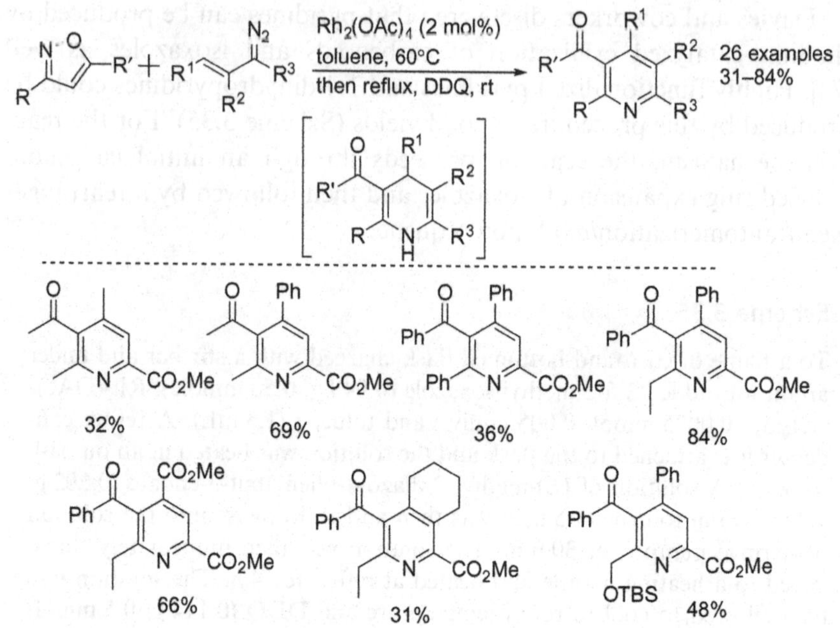

Scheme 3.35 *Rh-catalyzed synthesis of pyridines from carbenoids and isoxazoles.*

In 2010, Wang's group reported an ytterbium perfluorooctanoate [Yb (PFO)$_3$] catalyzed synthesis of 2-amino-3-cyanopyridine derivatives [78]. With aldehydes, ketones, malononitrile, and ammonium acetate as the substrates, 2-amino-3-cyanopyridines were prepared in a environmental benign manner in high yields and short reaction time (Scheme 3.36). For the reaction mechanism, the reaction started with the formation of enamine from ketone and ammonium acetate and the production of alkylidenemalononitrile by the condensation of aldehyde with malononitrile. Then under the assistance of ytterbium perfluorooctanoate, enamine reacted with alkylidenemalononitrile to give the key intermediate which will give the final product after cyclization and oxidation. More recently, using diamine functionalized [N-(2 amino ethyl)-3-amino propyl trimethoxy silane (AAPTMS)] mesoporous ZrO$_2$ (AAPTMS/m-ZrO$_2$) as the catalyst for this transformation was reported as well [79].

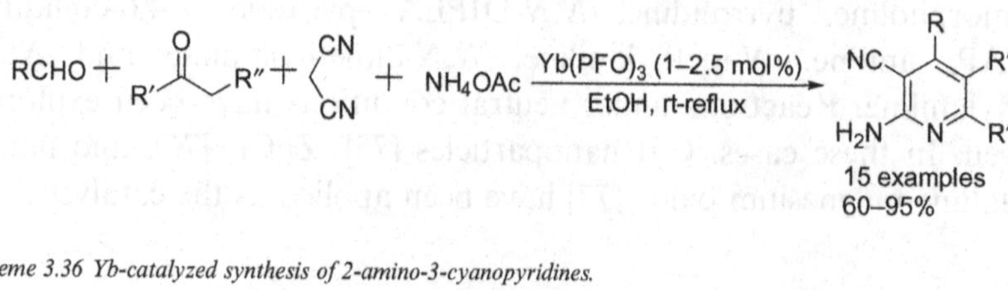

Scheme 3.36 *Yb-catalyzed synthesis of 2-amino-3-cyanopyridines.*

In 2014, Shang and coworkers reported $FeCl_3$-catalyzed four-component nucleophilic addition/intermolecular cyclization procedure for the construction of pyridines [80]. A broad range of polysubstituted pyridines was prepared in moderate to good yield from aryl aldehydes, β-keto esters, anilines, and malononitrile as simple and readily available starting materials (Scheme 3.37). The using of ethanol as reaction partner instead of anilines was studied and succeeded as well. More recently, the using of $SnCl_2 \cdot 2H_2O$ as the catalyst in water was also developed [81]. Good to excellent of the desired pyridines can be achieved.

Scheme 3.37

To a stirred solution of 4-nitrobenzaldehyde (0.151 g, 1 mmol), malononitrile (0.066 g, 1 mmol), ethyl acetoacetate (0.130 g, 1 mmol), and aniline (0.093 g, 1 mmol) in ethanol (3 mL) was added anhydrous $FeCl_3$ (0.080 g, 0.05 mmol). The mixture was heated in an oil bath at 70°C for 6 h and cooled to room temperature. Solvent was removed under vacuum, and the residue was purified by flash column chromatography on silica gel (200–300 mesh) with ethyl acetate and petroleum ether (1:10, v/v) as eluent to afford the desired product.

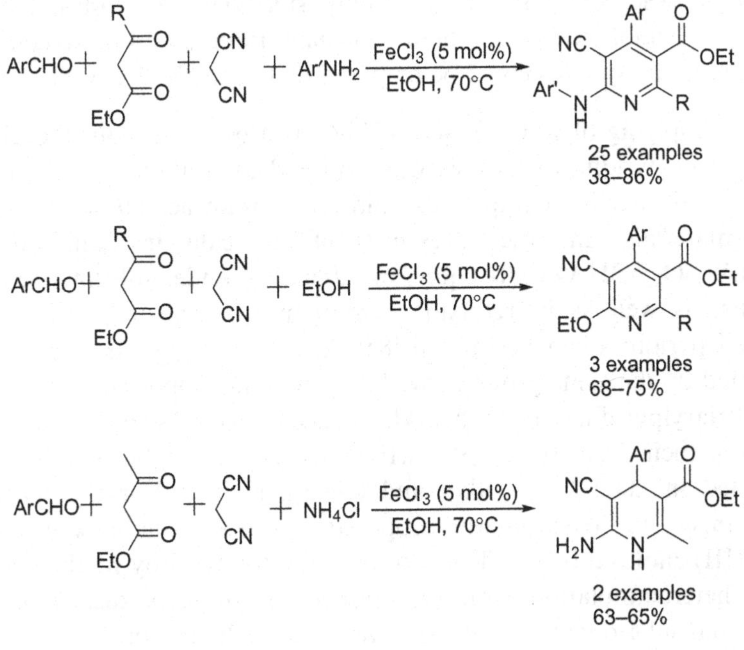

Scheme 3.37 Fe-catalyzed synthesis of 2-amino-3-cyanopyridines.

The using of chalcones, ketones, and ammonia as substrates with magnesium methoxide as the promoter to synthesis pyridines was described in 2013 [82]. Fair to good yields of the corresponding 2,4,6-triaryl pyridines have been formed in MeOH (Scheme 3.38). More recently, the combination of ceric ammonium nitrate (10 mol%) and pyrrolidine (20 mol%) as the catalytic system was explored for this transformation as well [83].

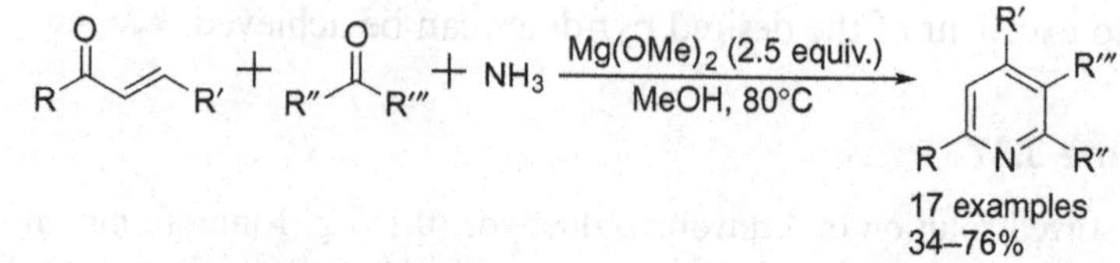

17 examples
34–76%

Scheme 3.38 Mg-promoted synthesis of pyridines.

An InCl$_3$-catalyzed sequential multicomponent reaction for pyridines synthesis was developed in 2014 [84]. Then on using 2-furfurylamine, β-dicarbonyl compounds, and α,β-unsaturated aldehydes as the substrates with ethanol as the solvent, good to excellent yields of highly substituted pyridines were prepared under the assistance of microwave irradiation (Scheme 3.39). 2-Furylmethyl side chain was lost as the byproduct. Quinolones, isoquinolines, phenanthridines, and more complex fused pyridine systems could be prepared by this method as well.

The applying of K$_5$CoW$_{12}$O$_{40}$ · 3H$_2$O as a heterogeneous recyclable catalyst in pyridines synthesis was explored as well. By using enaminones, β-dicarbonyl compounds, and ammonium acetate as the starting materials in the presence of a catalytic amount of K$_5$CoW$_{12}$O$_{40}$ · 3H$_2$O (0.01 equiv. or 1.0 mol%) under solvent free conditions, as well as in refluxing isopropanol, excellent yields of the desired pyridines can be formed [85]. Microwave-assisted version was reported by the same group as well [86]. A solid supported synthesis of 2,4,6-triarylpyridines from benzylideneacetophenones and urea, thiourea, or their derivatives with Bi(III) nitrate-Al$_2$O$_3$ as the catalyst was reported in 2005 by Razdan and coworkers [87]. For the reaction mechanism, the reaction was proposed to proceed via β-oxygenation of Bi(III)-enolized benzylideneacetophenone followed by Michael addition, heteroannulation with simultaneous *retro aldol* disproportionation and subsequent catalytic oxidation and dehydration.

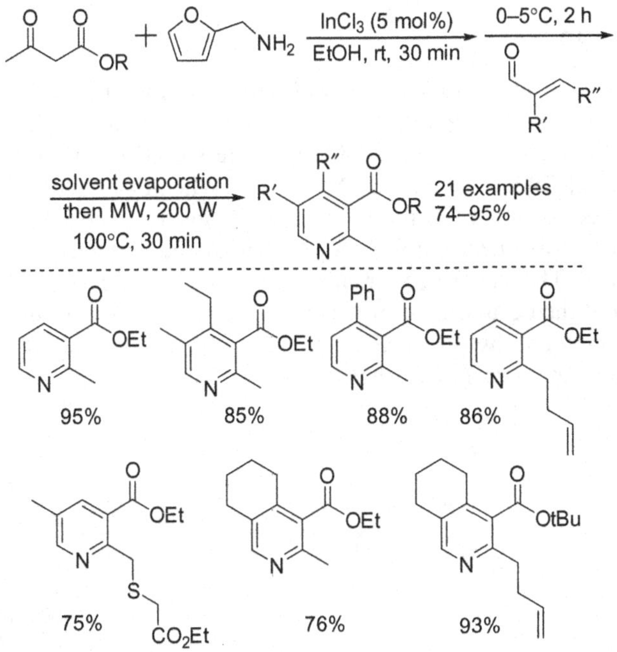

Scheme 3.39 In-catalyzed synthesis of pyridines with 2-furfurylamine as the amine source.

Jiang and coworkers developed a Cu-catalyzed three-component procedure for the synthesis of functionalized pyridines in 2015 [88]. Copper-catalyzed N−O bond cleavage/C−C/C−N bond formations were involved. Multi-substituted pyridines were produced in good yields from oxime acetates, activated methylene compounds, and a wide range of aldehydes bearing aryl, heteroaryl, vinyl, and trifluoromethyl groups (Scheme 3.40). For the reaction mechanism, the reaction started copper salts-catalyzed cleavage of the N−O bond of oxime acetates and generating copper enamide intermediate together with another Cu^{II} species. Subsequently, the nucleophilic addition of copper(II) enamide complex to 3-benzylidenepentane-2,4-dione (from benzaldehyde and acetylacetone) gives the imine intermediate. After tautomerization and intramolecular nucleophile attack affords the dihydropyridine complex which will give the final product after oxidation by Cu(II) species.

Scheme 3.40

The ketoxime acetates (0.3 mmol), aldehydes (0.3 mmol), activated methylene compound (1.5 equiv.), CuBr (10 mol%, 0.03 mmol, 4.26 mg), and Li$_2$CO$_3$ (20 mol%, 0.06 mmol, 4.44 mg) were stirred in DMSO (2.0 mL) at 120°C, in a 20 mL tube under N$_2$ for 6 h. When the reaction was completed (detected by TLC), the mixture was cooled to room temperature. The reaction was quenched with H$_2$O (10 mL) and extracted with EtOAc (3 × 10 mL). The combined organic layers were dried over anhydrous MgSO$_4$ and then evaporated in vacuum. The residue was purified by column chromatography on silica gel to afford the corresponding pyridines with hexanes/ethyl acetate as the eluent.

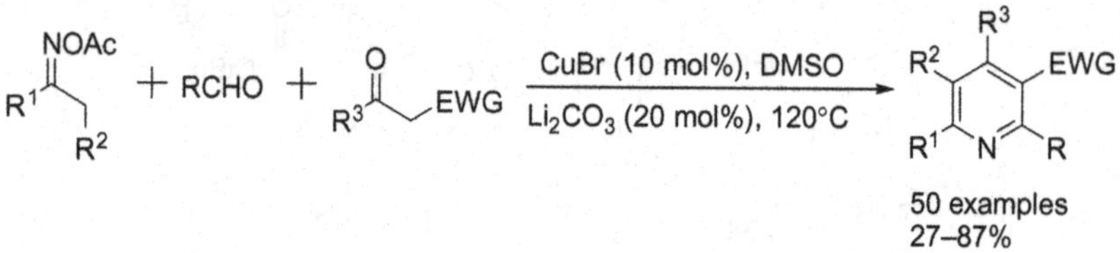

Scheme 3.40 Cu-catalyzed three-component procedure to synthesis pyridines.

In 2014, Odom and coworkers reported a titanium-catalyzed synthesis of 2-amino-3-cyanopyridines (Scheme 3.41) [89]. The reactions were performed in a one-pot manner with two manipulation steps. Firstly, titanium-catalyzed alkyne iminoamination and generation tautomers of 1,3-diimines. Then, 2-amino-3-cyanopyridines were formed in good to modest yields after treatment with base (DBU) and malononitrile. A Dimroth rearrangement mechanism for 2-aminopyridine formation was proposed based on the isolation of a 2-imino-1,2-dihydropyridine intermediate which undergoes rearrangement under the reaction conditions.

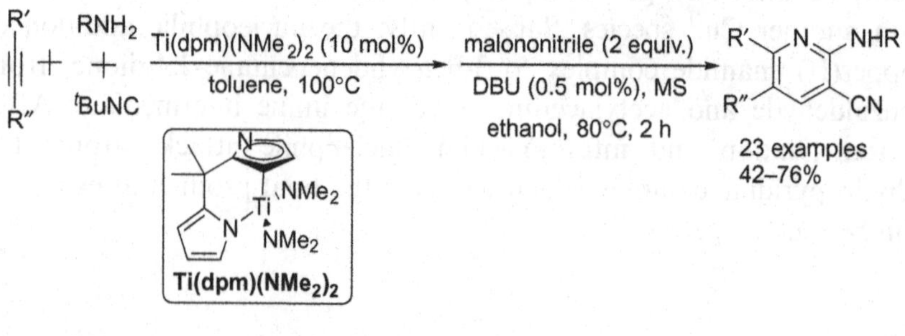

Scheme 3.41 Ti-catalyzed synthesis of 2-amino-3-cyanopyridines.

A Rh-catalyzed synthesis of pyridines from aldehydes, alkynes, and NH₄OAc via hydroacylation and *N*-annulation was reported by Jun and coworkers in 2012 [90]. More in detail, the reaction consisted of Rh(I)-catalyzed chelation-assisted hydroacylation of alkynes with aldehydes followed by Rh(III)-promoted *N*-annulation of the resulting α,β-enones with another alkyne and NH₃. Moderate to good yields of the desired pyridines were produced (Scheme 3.42).

Scheme 3.42

Chelation-assisted hydroacylation of 1-alkynes with aldehydes: To a 1 mL screw capped pressure vial (1 mL) was added isovaleraldehyde (17.2 mg, 0.20 mmol), 1-hexyne (32.9 mg, 0.40 mmol), (Ph₃P)₃RhCl (9.3 mg, 0.01 mmol), 2-amino-3-picoline (8.7 mg, 0.08 mmol), benzoic acid (4.9 mg, 0.04 mmol), and toluene (200 mg). The mixture was heated at 110°C for 4 h with stirring. After cooling the vessel to room temperature, the crude mixture was purified by column chromatography (*n*-hexane: Et₂O = 15:1) on silica gel to give 2-methyl-5- methylenenonan-4-one (25.9 mg, 77%) as a pale yellow oil.

N-Annulation of α,β-enone with internal alkyne: To a MeOH solution (0.2 mL) of 2-methyl-5-methylenenonan-4-one (33.7 mg, 0.20 mmol) and 4-octyne (44.1 mg, 0.40 mmol) were added [Cp*RhCl₂]₂ (3.1 mg, 0.005 mmol) and Cu(OAc)₂·H₂O (79.9 mg, 0.40 mmol), NH₄OAc (30.8 mg, 0.40 mmol), and the reaction mixture was stirred at 130°C under a nitrogen atmosphere for 6 h. After cooling to room temperature, the solvent was removed *in vacuo*, and the resulting crude mixture was subject to flash column chromatography (*n*-hexane:ethyl acetate = 90:1) to afford 3-butyl-2-isobutyl-5,6-dipropylpyridine (51.8 mg, 0.188 mmol) in 94% yield.

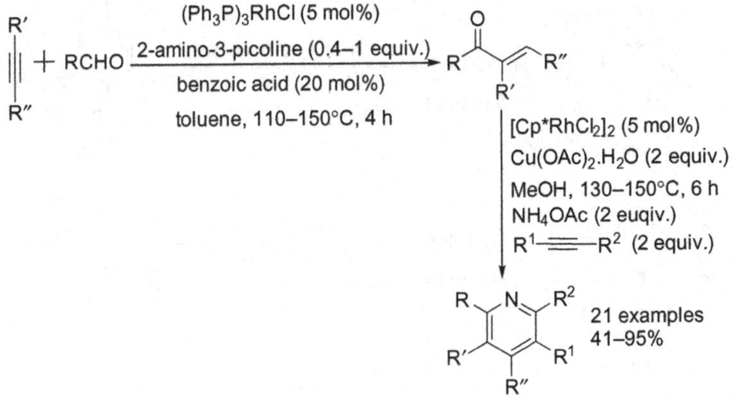

Scheme 3.42 Rh-catalyzed two-step synthesis of pyridines.

Hua and coworkers developed a rhodium-catalyzed pyridine derivatives preparation in 2012 [91]. By using aryl ketones, hydroxylamine, and alkynes as the substrates, via rhodium(III)-catalyzed C−H bond activation of the in situ generated aryl ketone oximes, and then cyclization with internal alkynes, good yields of the desired pyridines can be achieved (Scheme 3.43). Additionally, multisubstituted isoquinolines, γ-carbolines, furo[2,3-c]pyridines, thieno[2,3-c]pyridines, and benzofuro [2,3-c]pyridines can also be produced. In 2014, Zhang, Xiong, and coworkers further improved this methodology [92]. They succeeded to use ammonium acetate as the amine source for generation ketimines with ketones, and then reacted with alkynes to give the final products. In addition, water was applied as the solvent. Good yields can be achieved under atmosphere of oxygen at 100°C.

Scheme 3.43

To a 25 mL tube equipped with a magnetic stirrer, 1,2-diphenylacetylene (392.1 mg, 2.2 mmol), hydroxylamine hydrochloride (152.8 mg, 2.2 mmol), [Cp*RhCl$_2$]$_2$ (12.4 mg, 0.02 mmol), KOAc (412.2 mg, 4.2 mmol), acetophenone (240.4 mg, 2.0 mmol), and MeOH (10 mL) were added sequentially. The tube was sealed and stirred at 60°C in an oil bath for 18 h. After removal of the solvent under reduced pressure, purification was performed by flash column chromatography on silica gel with petroleum ether/ethyl acetate (gradient mixture ratio from 100:0 to 75:25) as eluant to afford isoquinoline.

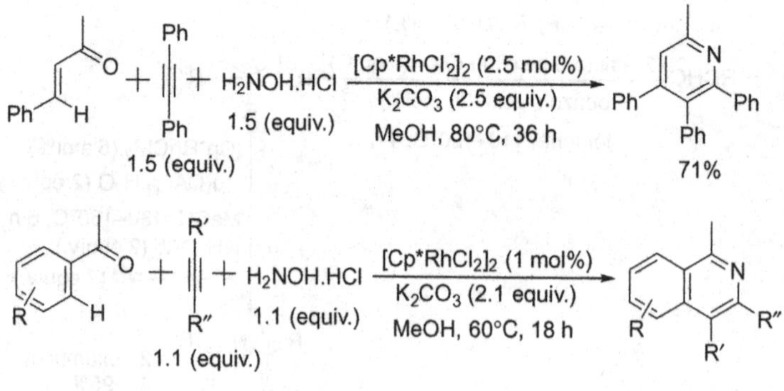

Scheme 3.43 Rh-catalyzed synthesis of pyridines from alkynes.

In 2014, Wan and coworkers developed a copper-catalyzed synthesis of pyridines from enaminones and ammonium chloride [93]. Under the assistant of copper catalyst and aerobic conditions, C—N and C=C bonds of enaminones were cleavage. 4-Unsubstituted pyridines were produced in good yields (Scheme 3.44).

Scheme 3.44

In a 25 mL round-bottom flask equipped with a stirring bar were located enaminone(s) (0.6 mmol) or (0.3 + 0.3 mmol, for the synthesis of unsymmetrical pyridines), ammonium chloride (0.3 mmol), CuI (0.06 mmol), and DMSO (2 mL). The vessel was then stirred at 120°C for 12 h at an air atmosphere. After completion of the reaction (TCL), the mixture was mixed with water (5 mL) and then extracted with ethyl acetate (3 × 10 mL). The combined organic phase was dried over anhydrous Na_2SO_4. After filtration, the organic solvent was removed from the solution under reduced pressure. The resulting residue was then subjected to silicon chromatography to give pure product by using mixed ethyl acetate/petroleum ether as eluent (v/v = 1/10).

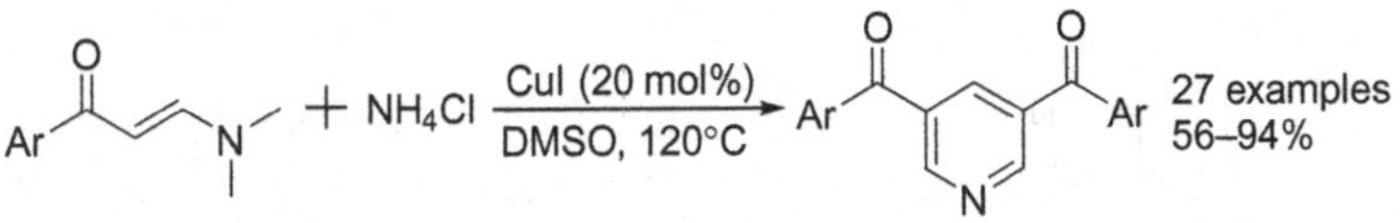

Scheme 3.44 Cu-catalyzed synthesis of pyridines from enaminones.

Guan and coworkers developed a copper-catalyzed coupling of oxime acetates with aldehydes for the synthesis of pyridines [94]. A wide range of functional groups can be tolerated and a variety of substituted pyridines were produced in good yields (Scheme 3.45). $NaHSO_3$ was discovered as a good inhibitor of hydrolysis of oxime acetates in this transformation. For the reaction mechanism, the reaction started with the cleavage of the N—O bond of the oxime acetate by Cu^+ gives the imine radical, which is supposed to react rapidly with another Cu^+ to give N-Cu^{2+} specie. Then its nucleophilic addition to benzaldehyde forms imine intermediate after self-tautomerization. Afterwards, condensation of the imine intermediate with a second oximeacetate which will give the final pyridines after electrocyclization and oxidation.

Scheme 3.45

In a 25 mL round bottom flask, the oxime acetates (0.9 mmol), aldehydes (0.3 mmol), CuBr (10 mol%, 4.3 mg), and NaHSO$_3$ (0.9 mmol, 93.6 mg) were stirred in DMSO (5.0 mL) at 120°C under Ar for 2.5 h. When the reaction was completed (detected by TLC), the mixture was cooled to room temperature. The reaction was quenched with H$_2$O (10 mL) and extracted with EtOAc (3 × 10 mL) or DCM (3 × 10 mL). The combined organic layers were dried over anhydrous Na$_2$SO$_4$ and then evaporated in vacuo. The residue was purified by column chromatography on silica gel to afford the corresponding pyridines with hexane/ethyl acetate as the eluent.

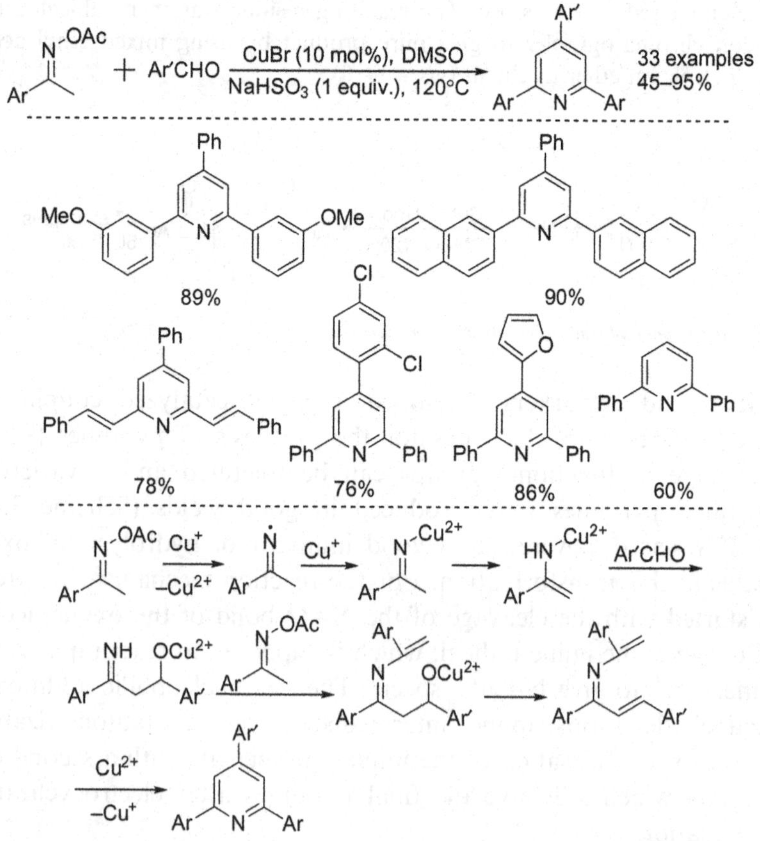

Scheme 3.45 Cu-catalyzed synthesis of pyridines from oxime acetates and aldehydes.

More recently, the same group reported another novel procedure for the synthesis of pyridines by cyclization of ketoxime acetates with ruthenium as the catalyst (Scheme 3.46) [95]. Here, a methyl carbon on DMF performed as a source of a one carbon synthon. For different propiophenone oxime carboxylates tested, such as propionate, *tert*-butyrate, and benzoate, showed similar reactivity as the corresponding acetate derivative. Only a trace of product was observed with pentafluorobenzoate.

Scheme 3.46

In a 25 mL round bottom flask, the ketoxime acetate (0.4 mmol), Ru (cod)Cl$_2$ (5 mol%, 5.6 mg), and NaHSO$_3$ (0.6 mmol, 62.4 mg) were stirred in DMF (2 mL) at 120°C in air. After completion of the reaction (detected by TLC), the reaction mixture was cooled to room temperature, extracted with ethyl acetate (20 mL), and washed with brine (20 mL). The organic layer was dried over by anhydrous Na$_2$SO$_4$ and evaporated in vacuo. The desired pyridine was obtained after purification by flash chromatography on silica gel with hexanes/ethyl acetate as the eluent.

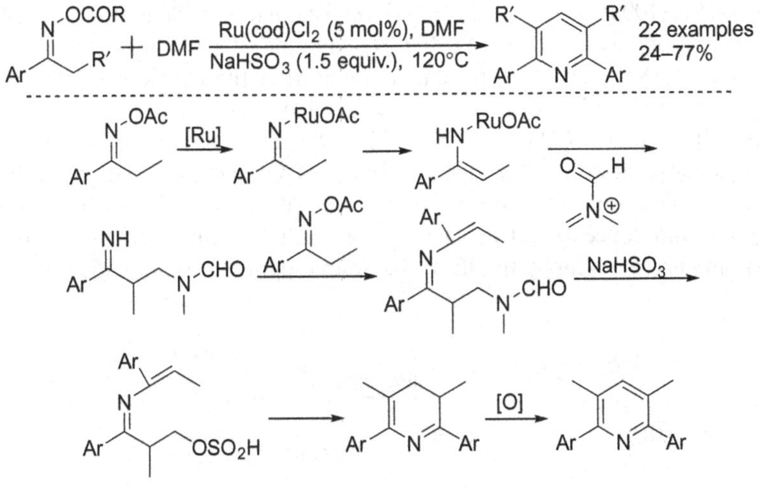

Scheme 3.46 Ru-catalyzed synthesis of pyridines from oxime acetates and DMF.

Müller's group developed a one-pot three-step four-component process for the construction of dihydropyrindines and tetrahydroquinolines [96]. Through a coupling-isomerization-Stork-enamine alkylation-cyclocondensation sequence of an electron poor (hetero)aryl halide, a terminal propargyl alcohol, a cyclic *N*-morpholino alkene and ammonium chloride, good yields of the desired products can be isolated [97].

In 2008, Katsumura and coworkers reported a novel methodology for the synthesis of 2-arylpyridines [98]. By utilizing the one-pot 6π-azaelectrocyclization and then treatment with a base, moderate to good yields of pyridine derivatives have been produced (Scheme 3.47).

Scheme 3.47

Method A: To a solution of methanesulfonamide (75 mg, 0.788 mmol), iodoolefin (100 mg, 0.394 mmol), and vinylstannane (0.788 mmol) in DMF (5 mL/mmol) were added Pd$_2$(dba)$_3$ (7 mg, 0.008 mmol), P(2-furyl)$_3$ (7 mg, 0.032 mmol) and, LiCl (34 mg, 0.788 mmol) at room temperature. After the reaction mixture was stirred at 80°C for 2–3 h, it was cooled to room temperature, and DBU (0.071 mL, 0.472 mmol) was added. The resulting mixture was stirred at this temperature for 1 h, quenched with H$_2$O, and extracted with ether. The organic layers were combined, washed with brine, dried over MgSO$_4$, filtered, and concentrated in vacuo. The residue was purified by silica gel column chromatography to afford the desired pyridine compounds.

Method B: To a solution of methanesulfonamide (75 mg, 0.788 mmol), iodoolefin (100 mg, 0.394 mmol) and vinylstannane (0.788 mmol) in DMF (5 mL/mmol) were added Pd(PhCN)$_2$ (15 mg, 0.039 mmol) and LiCl (34 mg, 0.788 mmol) at room temperature. After the reaction mixture was stirred at 50–70°C for 2–4 h, it was cooled to room temperature, and DBU (0.071 mL, 0.472 mmol) was added. The resulting mixture was stirred at this temperature for 1 h, quenched with H$_2$O, and extracted with ether. The organic layers were combined, washed with brine, dried over MgSO$_4$, filtered, and concentrated in vacuo. The residue was purified by silica gel column chromatography to afford the desired pyridine compounds.

[Pd]: Pd$_2$(dba)$_3$ (2 mol%), P(2-furyl)$_3$ (8 mol%), LiCl, DMF, 80°C
or Pd(PhCN)$_2$Cl$_2$ (10 mol%), LiCl, DMF, 50–70°C

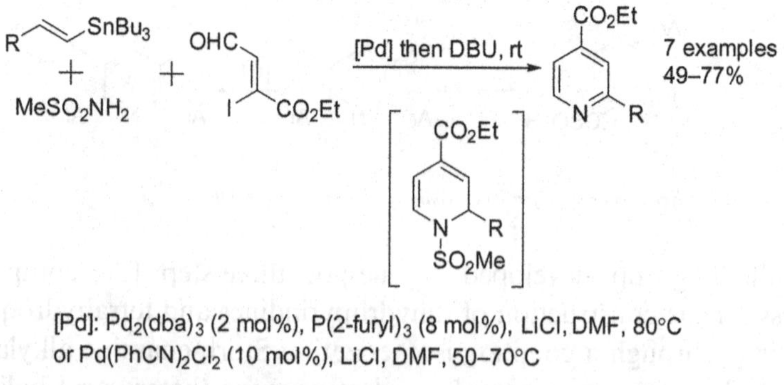

Scheme 3.47 Pd-catalyzed synthesis of pyridines through 6π-azaelectrocyclization.

3.5 SYNTHESIZED BY [4 + 2] CYCLIZATION REACTIONS

A rhodium-catalyzed one-pot synthesis of substituted pyridine derivatives from α,β-unsaturated ketoximes and alkynes was developed in 2008 by Cheng and coworkers [99]. Good yields of the desired pyridines can be obtained (Scheme 3.48). The reaction was proposed to proceed via rhodium-catalyzed chelation-assisted activation of the $\beta-C-H$ bond of α,β-unsaturated ketoximes and subsequent reaction with alkynes followed by reductive elimination, intramolecular electrocyclization, and aromatization to give highly substituted pyridine derivatives finally [100]. Later on, in their further studies, substituted isoquinolines and tetrahydroquinoline derivatives can be prepared by this catalyst system as well [101]. Their reaction mechanism was supported by isolation of the *ortho*-alkenylation products. Here, only asymmetric internal alkynes can be applied.

Scheme 3.48

A sealed tube containing RhCl(PPh$_3$)$_3$ (0.030 mmol, 3.0 mol%) was evacuated and purged with nitrogen gas three times. Freshly distilled toluene (2.0 mL), oxime (1.00 mmol), and alkyne (1.10 mmol) were sequentially added to the system and the reaction mixture was allowed to stir at 130°C for 3 h. The mixture was filtered through a short Celite pad and washed with dichloromethane several times. The filtrate was concentrated and the residue was purified on a neutral silica gel column using hexanes-ethyl acetate as eluent to afford the substituted pyridine derivative.

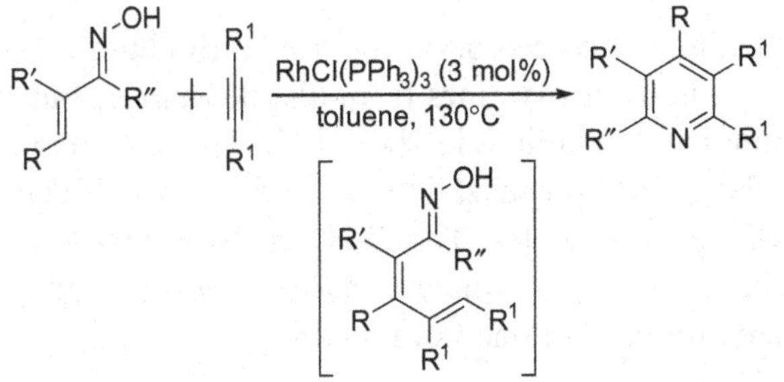

Scheme 3.48 Rh-catalyzed synthesis of pyridines from ketoximes and alkynes.

As the studies from Rovis and coworker, the problem of low regioselectivity with unsymmetrical alkynes can be solved by using [RhCp^{t}Cl$_2$]$_2$ or [RhCp*Cl$_2$]$_2$ (1.25 mol%) as the catalyst in the presence of K_2CO_3 (2 equiv.) in 2,2,2-trifluoroethanol at 45°C [102]. In 2012, Ellman, Bergman, and coworkers reported their achievements on synthesis of pyridines from ketoximes and terminal alkynes [103]. With triisopropyl phosphite as the ligand, substituted pyridines were formed in good yields with moderate to excellent regioselectivities (Scheme 3.49). More recently, they modified this methodology and applied in the synthesis of 3-fluoropyridines [104], with α-fluoro-α,β-unsaturated oximes and alkynes as the substrates. Oximes substituted with aryl, heteroaryl, and alkyl β-substituents were effectively reacted with both symmetrical and unsymmetrical alkynes with aryl and alkyl substituents.

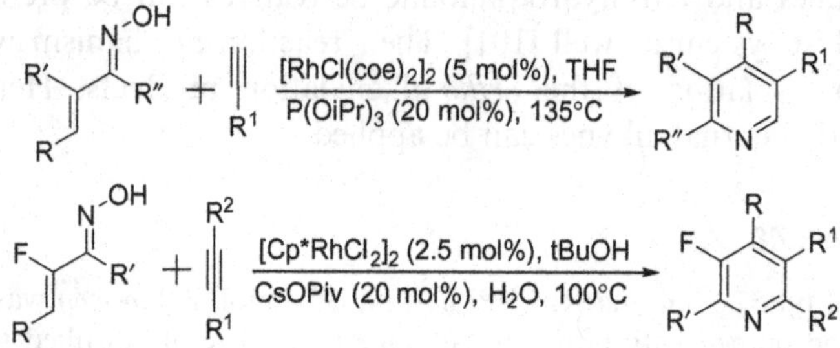

Scheme 3.49 Rh-catalyzed synthesis of 3-fluoropyridines.

This type of transformation was further studied by Huang and coworkers. They developed a Rh(III)-catalyzed procedure for the synthesis of azole-fused-pyridines by cyclization of azoles-based ketoximes with alkynes [105]. It was reported that this transformation can be performed under air as well [106].

Ellman, Bergman, and coworker reported a rhodium-catalyzed procedure for the synthesis of pyridines from alkynes and α,β-unsaturated N-benzyl aldimines and ketimines in 2008 [107]. The reaction proceeded via C-H alkenylation/electrocyclization/aromatization sequence through dihydropyridine intermediates. The C-H activated complex was isolated and determination by X-ray analysis. Good yields of highly substituted pyridines were produced in one-pot manner (Scheme 3.50).

Scheme 3.50

The desired alkyne (3.0 mmol) was placed in a sealable glass vessel in an inert atmosphere box. To this was added [RhCl(coe)$_2$]$_2$ (0.015 mmol) dissolved in 2 mL of toluene, followed by diethyl-(4-N,N-dimethylamino)-phenyl)phosphine ((p-DMAPh)PEt$_2$) (0.03 mmol) dissolved in 2 mL of toluene, followed by the desired imine (0.6 mmol) dissolved in 2 mL of toluene. The tube was then sealed, removed from the inert atmosphere box, and heated in a 100°C oil bath for the specified length of time. The tube was then allowed to cool to room temperature. Then 2,2,2-trifluoroethanol (2 mL, 25% v/v) and 10% Pd/C (20 theoretical wt%) were added. Reactions at room temperature were fitted with a drying tube, and those at 75°C were transferred to a round bottom flask and fitted with a reflux condenser. The solution was stirred while open to air at the specified temperature for the specified time. At room temperature, the drying tube or reflux condenser was removed and the flask was fitted with a hydrogen balloon. The flask was evacuated and back-filled with H$_2$ five times and the solution was stirred at room temperature for 12 h under a hydrogen atmosphere. The pure product can be obtained after purification.

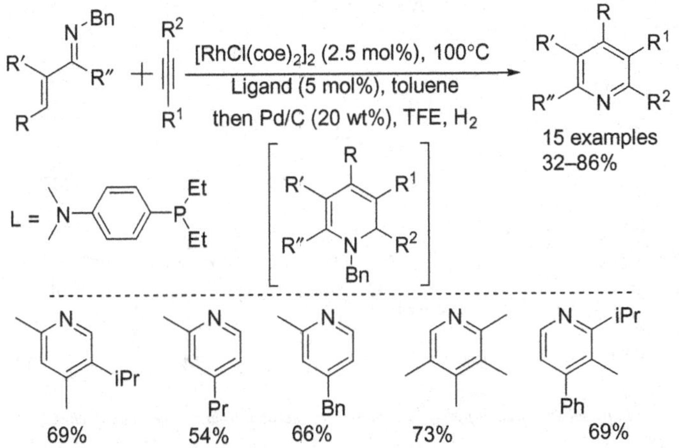

Scheme 3.50 Rh-catalyzed synthesis of pyridines from α,β-unsaturated N-benzyl aldimines and ketimines.

Rovis and Neely found that instead of alkynes, alkenes can be applied as substrates in pyridines synthesis with rhodium as the catalyst [108]. Good yields of pyridine derivatives were isolated from the corresponding α,β-unsaturated oxime esters and alkenes in general (Scheme 3.51). In the case of activated alkenes, exceptionally regioselective and high-yielding can be observed. Based on their mechanistic studies, the author proposed that heterocycle formation proceeds via reversible C−H activation, alkene insertion, and a C−N bond formation/N−O bond cleavage process.

Scheme 3.51

A 0.5 dram vial was charged with oxime ester (0.21 mmol) and AgOAc (73.6 mg, 2.1 equiv.) and a solution of [RhCp*Cl₂]₂ (3.3 mg, 0.025 equiv.) and alkene (0.252 mmol, 1.2 equiv.) in 0.7 mL 2:1 DCE/AcOH was added. The vial was flushed with argon, sealed and heated at 85°C in an aluminum heating block for 14 h. The solids were filtered and the mixture was diluted with DCM and washed with 15% Na₂CO₃. The aqueous layer was extracted twice with DCM and the combined organic layers were dried over MgSO₄, filtered and concentrated in vacuo. The crude product was purified by flash column chromatography.

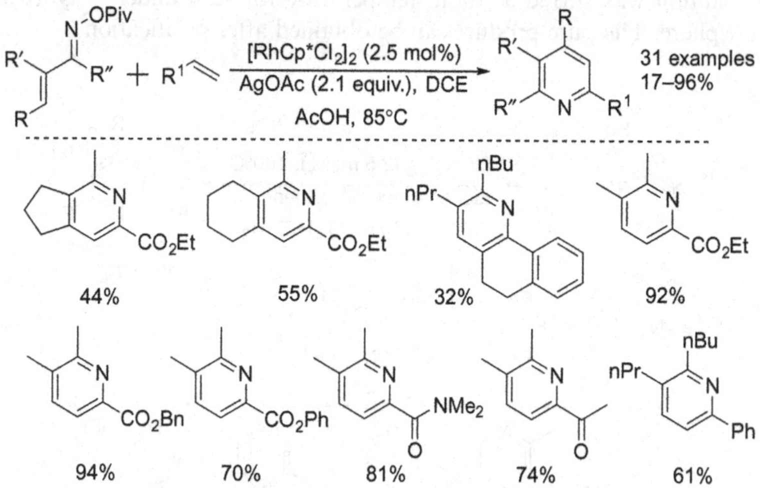

Scheme 3.51 Rh-catalyzed synthesis of pyridines from α,β-unsaturated oxime esters and alkenes.

Allyl amines and alkynes were explored as starting materials for pyridines synthesis by Jun and coworkers as well [109]. The reaction proceeded through a sequential Cu(II)-promoted dehydrogenation of the allylamine and Rh(III)-catalyzed *N*-annulation of the resulting α,β-unsaturated imine and alkyne. Moderate to good yields of pyridines can be isolated (Scheme 3.52). This transformation was later on explored with ruthenium catalyst [110]. In the presence of [{RuCl₂(*p*-cymene)}₂] (0.1 equiv.), KPF₆ (0.1 equiv.), and Cu(OAc)₂ (1 equiv.) in tAmOH at 100°C, the desired pyridine derivatives were formed in good yields. In this case, the reaction started with C-H activation and then insertion to alkynes which is different from the rhodium catalyzed case.

Scheme 3.52

A 1 mL pressure vial was charged with 2-methallylamine (14 mg (0.2 mmol)), 4-octyne (26.4 mg, (0.24 mmol)), [Cp*RhCl$_2$]$_2$ (1.5 mg (0.00625 mmol)), Cu (OAc)$_2$·H$_2$O (80 mg (0.4 mmol)), and methanol (200 μL). The reaction mixture was stirred at 100°C preheated oil-bath for 4 h. After cooling the vessel to room temperature, the crude mixture was purified by column chromatography (n-hexane:ethyl acetate = 2:1) on silica gel to give 5-methyl-2,3-dipropyl-pyridine (81.6 mg, 92%) as a pale yellow oil.

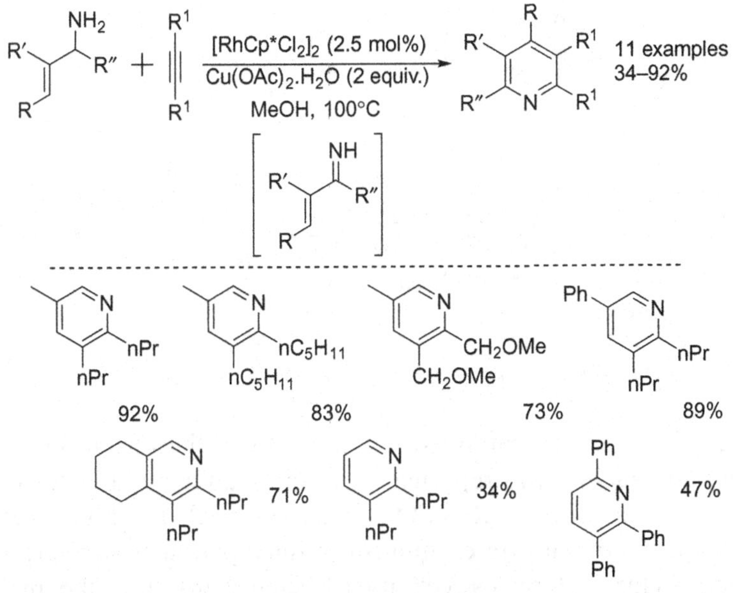

Scheme 3.52 Rh-catalyzed synthesis of pyridines from allyl amines and alkynes.

More recently, Dong and coworkers developed a novel methodology for the construction of highly functionalized pyridines from N-sulfonyl ketimines and alkynes [111]. Rhodium was applied as the catalyst and good yields can be achieved (Scheme 3.53). The N−S bond plays as an internal oxidant, C−C/C−N bond formation and S−N/S−C (or S−O) bond cleavage, along with desulfonylation were involved.

Scheme 3.53

(*E*)-3-Styrylbenzo[d]isothiazole 1,1-dioxide (26.9 mg, 0.1 mmol), diphenyl acetylene (20.4 mg, 0.12 mmol), [Cp*RhCl₂]₂ (1.6 mg, 2.5 mol%), AgBF₄ (5 mg, 0.05 mmol), and HOAc (30 μL, 0.5 mmol) were stirred in DCE (1.0 mL) under Ar at 60°C for 12 h. After completion, the reaction mixture was cooled down to room temperature and then neutralized with K₂CO₃. The reaction mixture was purified by flash chromatography eluting with ethyl acetate and petroleum ether (1:100) to give the pure product.

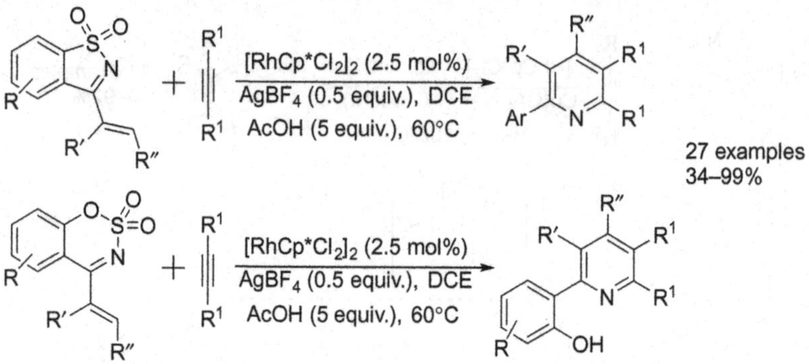

Scheme 3.53 Rh-catalyzed synthesis of pyridines from N-sulfonyl ketimines and alkynes.

In 2012, a rhenium-catalyzed synthesis of pyridines from β-enamino ketones and alkynes was reported by Takai, Kuninobu, and coworkers [112]. In the presence of Re₂(CO)₁₀, with *N*-acetyl β-enamino ketones and alkynes as the substrates, multisubstituted pyridines were produced in good yields and regioselectivity (Scheme 3.54). For the reaction mechanism, the reaction proceeded via insertion of alkynes into a carbon−carbon single bond of β-enamino ketones, intramolecular nucleophilic cyclization, and elimination of acetic acid. And the presences of *N*-acetyl moieties are important for the reaction selectivity.

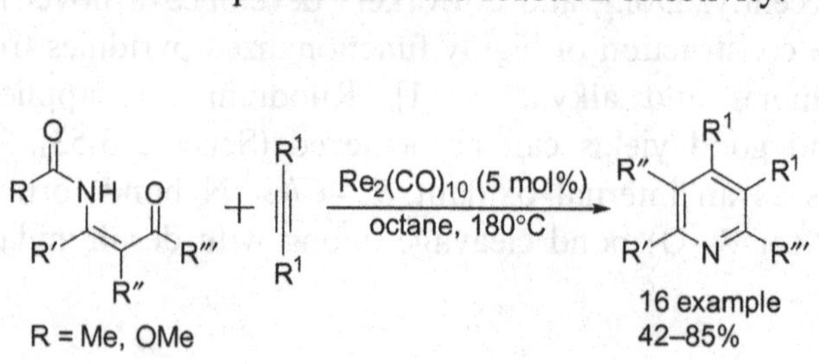

Scheme 3.54 Re-catalyzed synthesis of pyridines from β-enamino ketones and alkynes.

Recently, Wang, Yu, and coworkers developed a ruthenium-catalyzed cycloaddition of enamides and alkynes for the synthesis of pyridine derivatives [113]. Broad substrate scope, high efficiency, good functional group tolerance, and excellent regioselectivities were observed for this reaction (Scheme 3.55). Regarding the reaction mechanism, density functional theory (DFT) calculations and experiments have been carried out. Based on DFT calculations, it was suggested that this reaction started with a concerted metalation deprotonation of the enamide by the acetate group in the Ru catalyst, which generates a six-membered ruthenacycle intermediate. Then alkyne inserts into the Ru—C bond of the six-membered ruthenacycle, giving rise to an eight-membered ruthenacycle intermediate. The carbonyl group (which comes originally from the enamide substrate and is coordinated to the Ru center in the eight-membered ruthenacycle intermediate) then inserts into the Ru—C bond to give an intermediate, which produces the final pyridine product through further dehydration. Alkyne insertion step is a regio-determining step and prefers to have the aryl groups of the used alkynes stay away from the catalyst in order to avoid repulsion of aryl group with the enamide moiety in the six-membered ruthenacycle and to keep the conjugation between the aryl group and the triple C—C bond of the alkynes. Hence, the aryl groups of the applied alkynes are regioselectivity stay in the β-position of the final pyridines.

Scheme 3.55

To a Schlenk tube were added enamides (0.3 mmol), alkynes (0.45–0.60 mmol), [RuCl$_2$(p-cymene)]$_2$ (5 mol%), Na$_2$CO$_3$ (0.3 mmol), KOAc (0.3 mmol), and toluene (2.0 mL). Then the tube was charged with argon, and was stirred at 100°C for the indicated time (24–48 h) until complete consumption of starting material as monitored by TLC and GC-MS analysis. After the reaction was finished, the reaction mixture was washed with brine. The aqueous phase was extracted with ethyl acetate. The combined organic extracts were dried over Na$_2$SO$_4$, concentrated in vacuum, and the resulting residue was purified by silica gel column chromatography (hexane/ethyl acetate) to afford the desired product.

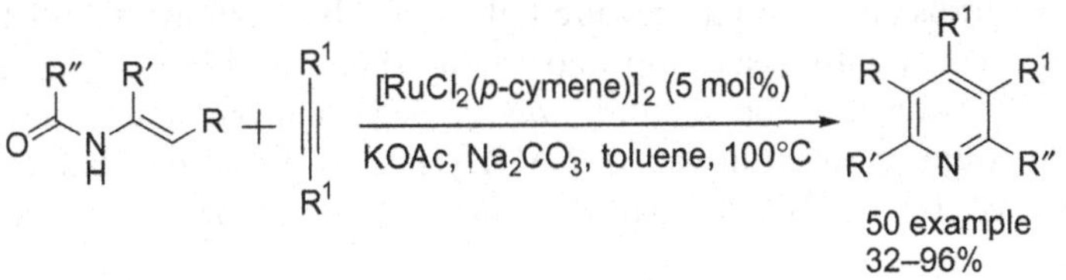

50 example
32–96%

Scheme 3.55 Ru-catalyzed synthesis of pyridines from enamides and alkynes.

In 2008, a gold-catalyzed cycloaddition of captodative dienynes with nitriles to provide pyridines was developed [114]. The reaction proceeded via intermolecular hetero-dehydro-diels-alder cycloaddition and gave the desired pyridines in good yields regioselectively (Scheme 3.56).

Scheme 3.56

AuClPEt₃ (5 mol%, 9 mg) and AgSbF₆ (5 mol%, 9 mg) were added to a solution of the corresponding dienyne (0.5 mmol) and the appropriate nitrile (20 equiv., 10 mmol) in dry 1,2-dichloroethane (5 mL). The reaction mixture was stirred at 85°C until complete disappearance of dyenine was observed by TLC or GC/MS. Solvent was removed under reduced pressure and the crude mixture was purified by flash chromatography on silica gel using mixtures of hexane and AcOEt.

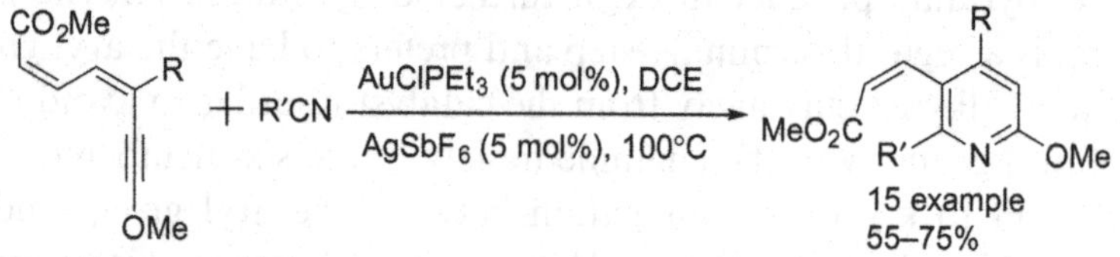

Scheme 3.56 Au-catalyzed synthesis of pyridines from dienynes with nitriles.

Ogoshi and coworkers reported a nickel-catalyzed dehydrogenative cycloaddition of 1,3-dienes and nitriles to give pyridine derivatives in 2011 [115]. Moderate to good yields of the pyridines were formed (Scheme 3.57). In addition to usual nitriles, di- and tricyano compounds can be applied as well and give the corresponding polypyridine derivatives in good yields. Notably, hydrogen molecular was the only waster; no extra oxidant was required here.

Scheme 3.57

To a solution of Ni(cod)₂ (10 mol%, relative to nitrile), PCy₃ (40 mol%) and nitrile in toluene was added 1,3-diene (4.0 equiv.). The resulting mixture was transferred into a pressure-tight test-tube (volume: 8.5 mL). The test-tube was tightly sealed up and thermostated at 130°C for a giving time. After the reaction mixture was cooled to room temperature, all volatiles were removed in vacuo. NMR yields of the crude product were determined by ^{1}H NMR analysis using 1,3,5-trioxane, 1,4-dioxane or nitromethane as an internal standard. The residue was purified by silica gel column chromatography (eluent: hexane, silica gel: Wakogel® 50NH₂) to afford the desired product.

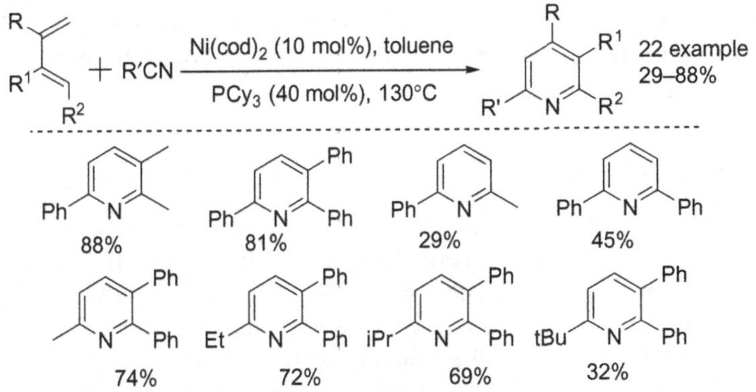

Scheme 3.57 Ni-catalyzed synthesis of pyridines from dienes with nitriles.

The reaction of ethyl acetoacetate with α,β-unsaturated oximes in the presence of FeCl₃ (5 mol%) via Michael addition was found to give pyridine derivative as well [116]. However, relative high reaction temperature was required (150–160°C).

In 2015, Donohoe and coworkers discovered that pyridines can also be prepared by palladium-catalyzed enolate α-alkenylation of ketones [117]. In this methodology, the reaction started with a protected β-haloalkenylaldehyde which then go α-alkenylation with a ketone to afford a 1,5-dicarbonyl surrogate, which then undergoes cyclization/double elimination to the corresponding pyridine product. The β-haloalkenylaldehyde starting materials can be obtained from the corresponding methylene ketone via Vilsmeier haloformylation. Moderate yields of the pyridine derivatives were isolated (Scheme 3.58).

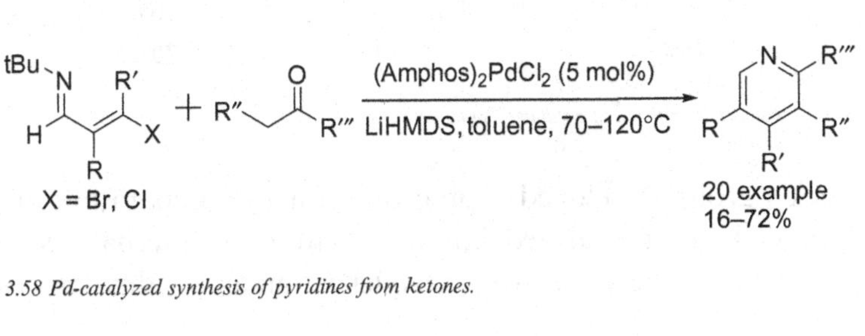

Scheme 3.58 Pd-catalyzed synthesis of pyridines from ketones.

Liebeskind and Liu developed a copper-catalyzed synthesis of pyridines from α,β-unsaturated ketoxime O-pentafluorobenzoates and alkenylboronic acids in 2008 [118]. This cascade reaction consisted of a novel N-iminative, Cu-catalyzed cross-coupling of alkenylboronic acids at the NsO bond of α,β-unsaturated ketoxime O-pentafluorobenzoates, electrocyclization of the resulting 3-azatriene, and air oxidation at the last stage. Highly substituted pyridines were formed in moderate to excellent isolated yields (Scheme 3.59).

Scheme 3.59

Dry DMF (2 mL) was added to a Schlenk tube containing the α,β-unsaturated ketoxime O-pentafluorobenzoate (0.1 mmol), the boronic acid (0.12 mmol), 4 Å molecular sieves (100 mg), and Cu(OAc)$_2$ (1.8 mg, 0.01 mmol). The reaction mixture was stirred under air at 50°C for 2 h and then at 90°C for 3 h. The reaction mixture was extracted with ether and the organic phase was washed with brine and then dried over MgSO$_4$. After evaporation of the solvent the residue was subjected to flash chromatography giving the corresponding pyridine.

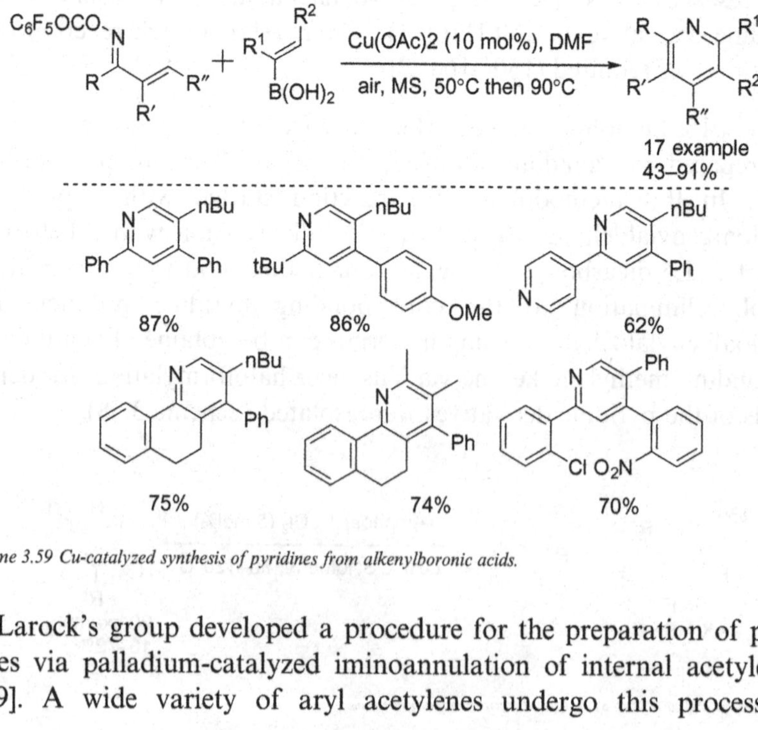

Scheme 3.59 Cu-catalyzed synthesis of pyridines from alkenylboronic acids.

Larock's group developed a procedure for the preparation of pyridines via palladium-catalyzed iminoannulation of internal acetylenes [119]. A wide variety of aryl acetylenes undergo this process in moderate to excellent yields, with high regioselectivity (Scheme 3.60). In the case of using imine of *o*-iodobenzaldehyde as the substrates, isoquinolines can be produced. The cyclization of the imines with allenes to build pyridine cores was reported by Frühauf and coworkers in 1999 [120].

Scheme 3.60

DMF (10 mL), Pd(OAc)$_2$ (6 mg, 0.027 mmol), PPh$_3$ (13 mg, 0.05 mmol), Na$_2$CO$_3$ (53 mg, 0.5 mmol), and the alkyne (1.0 mmol) were placed in a 4 dram vial. The contents were mixed and the appropriate imine (0.5 mmol) was added. The vial was flushed with nitrogen and heated in an oil bath at 100°C for the indicated period of time. The reaction was monitored by TLC to establish completion. The reaction mixture was then cooled to room temperature, diluted with 30 mL of ether, washed with 45 mL of saturated NH$_4$Cl, dried (Na$_2$SO$_4$), and filtered. The solvent was evaporated under reduced pressure, and the product was isolated by chromatography on a silica gel column.

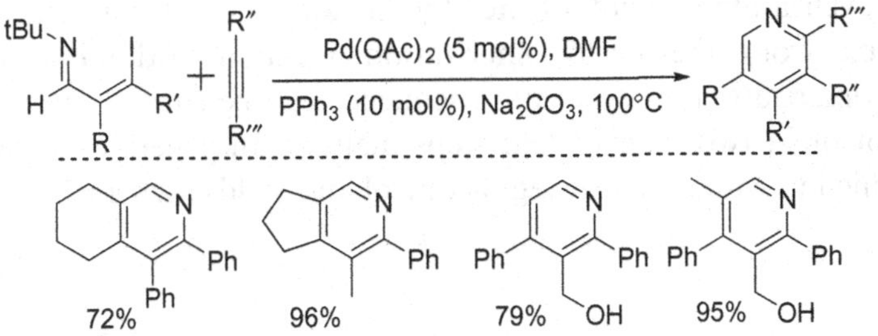

Scheme 3.60 Pd-catalyzed synthesis of pyridines from imines and alkynes.

3.6 SYNTHESIZED BY [5 + 1] CYCLIZATION REACTIONS

Micalizio and coworker reported a three-component coupling sequence for the synthesis of substituted pyridines in 2012 [121]. The reaction proceeded through nucleophilic addition of a dithiane anion to an α,β-unsaturated carbonyl followed by metallacycle-mediated union of the resulting allylic alcohol with preformed trimethylsilane-imines (generated in situ by the low temperature reaction of lithium hexamethyldisilazide with an aldehyde) and Ag(I)- or Hg(II)-mediated ring closure. Good yields of substituted pyridines were isolated.

A novel strategy for the synthesis of pyridines from aldehyde, enamide, and isonitrile was described by Wang and coworkers in 2013 [122]. The reaction works under mild reaction conditions and good to excellent yields can be achieved. Mechanistically, this cascade reaction consisted by $Zn(OTf)_2$-promoted [1 + 5] cycloaddition of isonitrile with N-formylmethyl-substituted enamide, facile aerobic oxidative aromatization and intermolecular acyl transfer from the pyridinium nitrogen to the 5-hydroxy oxygen, and finally acylation of the 4-amino group by an external acyl chloride efficiently afforded 2-substituted 4-acylamino-5-acyloxypyridines.

Trost's group reported a ruthenium-catalyzed method for the synthesis of substituted pyridines in 2007 [123]. The reaction started with $[CpRu(CH_3CN)_3]PF_6$-catalyzed cyclization of primary and secondary propargyl diynols into the corresponding unsaturated ketones and aldehydes. Then the unsaturated ketones and aldehydes were converted into 1-azatrienes which in turn undergo a subsequent electrocyclization-dehydration to provide the desired pyridines. A broad range of functional groups can be tolerated in this methodology. The yield of the electrocyclization was not affected by the double-bond isomeric mixture obtained from the cycloisomerization reaction. Additionally, this cycloisomerization-6π-cyclization sequence can be carried out either in one pot or in two independent steps manner. In general, the reaction performed in with excellent regio-control and yields (Scheme 3.61).

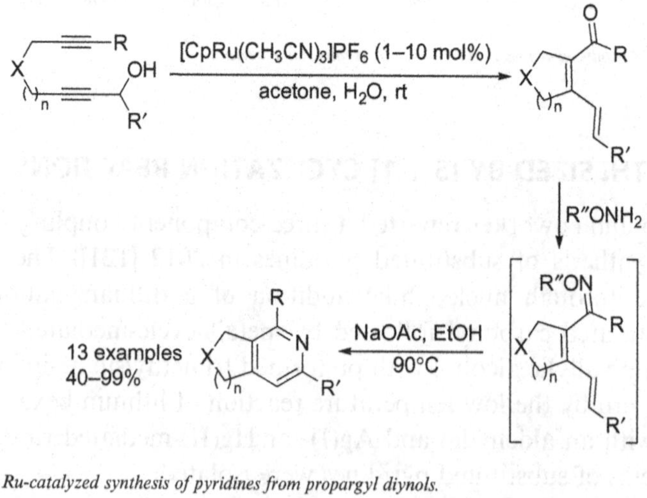

Scheme 3.61 Ru-catalyzed synthesis of pyridines from propargyl diynols.

Yudin and coworkers reported a palladium-catalyzed synthesis of multisubstituted pyridines from amino allenes, aldehydes, and aryl iodides in 2013 [124]. Moderate to good yields of the desired pyridines were prepared from readily available building blocks (Scheme 3.62).

Scheme 3.62

Step 1: To an oven-dried small vial containing a solution of α-amino allene (1.0 mmol) in anhydrous DCM (2 mL) was added anhydrous MgSO$_4$ (100 mg) and aldehyde (1.0 mmol). The reaction was stirred at room temperature for approximately 5–8 h, at which point the complete conversion of starting material to imine was confirmed by crude ^{1}H NMR. The resulting mixture was filtered through celite and concentrated to obtain the "skipped" allenyl imine as an orange oil which was carried over to the next step without further purification.

Step 2: To a flame-dried round-bottom flask equipped with rubber septum and N$_2$ line was added anhydrous DMF (2 mL), "skipped" allenyl imine (0.689 mmol), and aryl iodide (0.758 mmol). N$_2$ was bubbled through the solution for 30 min to eliminate O$_2$. Pd(PPh$_3$)$_4$ (0.069 mmol), and NaOAc (2.067 mmol) were then added. The reaction was stirred at 80°C for 12 h under N$_2$ then exposed to air and stirred for another 12 h, maintaining a temperature of 80°C. Subsequently, the mixture was cooled to room temperature. 10 mL of H$_2$O was added and the product was extracted with Et$_2$O (10 mL × 3). The organic layer was concentrated to yield a crude mixture which was purified via flash column chromatography on silica gel with 0 to 10% EtOAc in hexanes gradient elution system to obtain the pyridine products.

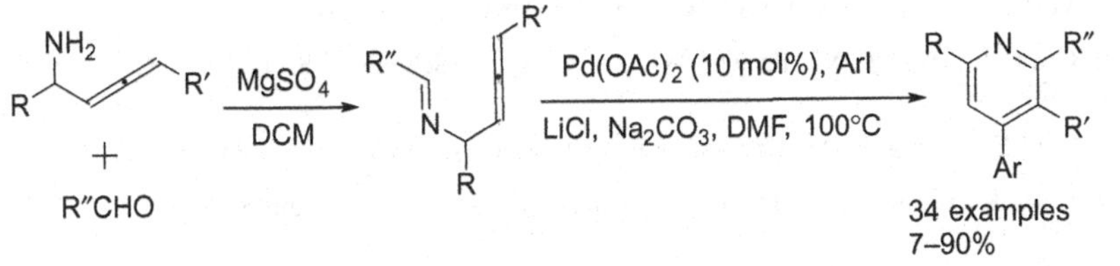

Scheme 3.62 Pd-catalyzed synthesis of pyridines from allenes.

REFERENCES

[1] a. Takahashi, T.; Tsai, F.-Y.; Kotora, M. *J. Am. Chem. Soc.* **2000**, *122*, 4994–4995.
     b. Takahashi, T.; Tsai, F.-Y.; Li, Y.; Wang, H.; Kondo, Y.; Yamanaka, M., et al. *J. Am. Chem. Soc.* **2002**, *124*, 5059–5067.

[2] a. Suzuki, D.; Tanaka, R.; Urabe, H.; Sato, F. *J. Am. Chem. Soc.* **2002**, *124*, 3518–3519.

b. Tanaka, R.; Yuza, A.; Watai, Y.; Suzuki, D.; Takayama, Y.; Sato, F., et al. *J. Am. Chem. Soc.* **2005**, *127*, 7774–7780.
c. Suzuki, D.; Nobe, Y.; Watai, Y.; Tanaka, R.; Takayama, Y.; Sato, F., et al. *J. Am. Chem. Soc.* **2005**, *127*, 7474–7479.

[3] Ferré, K.; Toupet, L.; Guerchais, V. *Organometallics* **2002**, *21*, 2578–2580.

[4] Wakatsuki, Y.; Yamazaki, H. *J. C. S. Dalton* **1978**, 1278–1282.

[5] Naiman.; Voollhardt, K. P. C. *Angew. Chem. Int. Ed. Engl.* **1977**, *16*, 708–709.

[6] a. Heller, B.; Heller, D.; Wagler, P.; Oehme, G. *J. Mol. Catal. A: Chem.* **1998**, *136*, 219–233.
b. Heller, B.; Sundermann, B.; Fischer, C.; You, J.; Chen, W.; Drexler, H.-J., et al. *J. Org. Chem.* **2003**, *68*, 9221–9225.
c. Gutnov, A.; Heller, B.; Fischer, C.; Drexler, H.-J.; Spannenberg, A.; Sundermann, B., et al. *Angew. Chem. Int. Ed.* **2004**, *43*, 3795–3797.
d. Heller, B.; Gutnov, A.; Fischer, C.; Drexler, H.-J.; Spannenberg, A.; Redkin, D., et al. *Chem. Eur. J.* **2007**, *13*, 1117–1128.
e. Hapke, M.; Kral, K.; Fischer, C.; Spannenberg, A.; Gutnov, A.; Redkin, D., et al. *J. Org. Chem.* **2010**, *75*, 3993–4003.

[7] Hoshi, T.; Katano, M.; Nozawa, E.; Suzukia, T.; Hagiwara, H. *Tetrahedron Lett.* **2004**, *45*, 3489–3491.

[8] Turek, P.; Hocek, M.; Pohl, R.; Klepetárová, B.; Kotora, M. *Eur. J. Org. Chem.* **2008**, 3335–3343.

[9] a. Senaiar, R. S.; Young, D. D.; Deiters, A. *Chem. Commun.* **2006**, 1313–1315.
b. Young, D. D.; Deiters, A. *Angew. Chem. Int. Ed.* **2007**, *46*, 5187–5190.

[10] Thiel.; Spannenberg, A.; Hapke, M. *ChemCatChem* **2013**, *5*, 2865–2868.

[11] Thiel.; Hapke, M. *J. Mol. Catal. A: Chem.* **2014**, *383–384*, 153–158.

[12] Geny, A.; Agenet, N.; Iannazzo, L.; Malacria, M.; Aubert, C.; Gandon, V. *Angew. Chem. Int. Ed.* **2009**, *48*, 1810–1813.

[13] Garcia, P.; Evanno, Y.; George, P.; Sevrin, M.; Ricci, G.; Malacria, M., et al. *Chem. Eur. J.* **2012**, *18*, 4337–4344.

[14] Garcia, P.; Evanno, Y.; George, P.; Sevrin, M.; Ricci, G.; Malacria, M., et al. *Org. Lett.* **2011**, *13*, 2030–2033.

[15] Varela, J. A.; Castedo, L.; Saá, C. *Org. Lett.* **1999**, *1*, 2141–2143.

[16] Gray, L.; Wang, X.; Brown, W. C.; Kuai, L.; Schreiber, S. L. *Org. Lett.* **2008**, *10*, 2621–2624.

[17] Zou, Y.; Young, D. D.; Cruz-Montanez, A.; Deiters, A. *Org. Lett.* **2008**, *10*, 4661–4664.

[18] Kase, K.; Goswami, A.; Ohtaki, K.; Tanabe, E.; Saino, N.; Okamoto, S. *Org. Lett.* **2007**, *9*, 931–934.

[19] Sugiyama, Y.-k; Okamoto, S. *J. Polym. Sci., Part A: Polym. Chem.* **2015**. http://dx.doi.org/pola.27780.

[20] Nicolaus, N.; Schmalz, H.-G. *Synlett* **2010**, 2071–2074.

[21] Saá, C.; Crotts, D. D.; Hsu, G.; Vollhardt, K. P. C. *Synlett* **1994**, 487–489.

[22] a. Yuan, C.; Chang, C.-T.; Axelrod, A.; Siegel, D. *J. Am. Chem. Soc.* **2010**, *132*, 5924–5925.
b. Yuan, C.; Chang, C.-T.; Siegel, D. *J. Org. Chem.* **2013**, *78*, 5647–5668.

[23] Varela, J. A.; Castedo, L.; Saá, C. *J. Org. Chem.* **1997**, *62*, 4189–4192.

[24] Varela, J. A.; Castedo, L.; Saá, C. *J. Am. Chem. Soc.* **1998**, *120*, 12147–12148.

[25] Goswami, A.; Ohtaki, K.; Kase, K.; Ito, T.; Okamoto, S. *Adv. Synth. Catal.* **2008**, *350*, 143–152.

[26] Weding, N.; Jackstell, R.; Jiao, H.; Spannenberg, A.; Hapke, M. *Adv. Synth. Catal.* **2011**, *353*, 3423–3433.

[27] Boñaga, L. V. R.; Zhang, H.-C.; Moretto, A. F.; Ye, H.; Gauthier, D. A.; Li, J., et al. *J. Am. Chem. Soc.* **2005**, *127*, 3473–3485.

[28] Yamamoto, Y.; Ogawa, R.; Itoh, K. *J. Am. Chem. Soc.* **2001**, *123*, 6189–6190.

[29] Yamamoto, Y.; Okuda, S.; Itoh, K. *Chem. Commun.* **2001**, 1102–1103.

[30] Yamamoto, Y.; Kinpara, K.; Saigoku, T.; Takagishi, H.; Okuda, S.; Nishiyama, H., et al. *J. Am. Chem. Soc.* **2005**, *127*, 605–613.

[31] Yamamoto, Y.; Kinpara, K.; Nishiyama, H.; Itoh, K. *Adv. Synth. Catal.* **2005**, *347*, 1913–1916.

[32] Yamamoto, Y.; Kinpara, K.; Ogawa, R.; Nishiyama, H.; Itoh, K. *Chem. Eur. J.* **2006**, *12*, 5618–5631.

[33] Varela, J. A.; Castedo, L.; Saá, C. *J. Org. Chem.* **2003**, *68*, 8595–8598.

[34] Medina, S.; Domínguez, G.; Pérez-Castells, J. *Org. Lett.* **2012**, *14*, 4982–4985.

[35] Xu, F.; Wang, C.; Li, X.; Wan, B. *ChemSusChem* **2012**, *5*, 854–857.

[36] Nissen, F.; Richard, V.; Alayrac, C.; Witulski, B. *Chem. Commun.* **2011**, *47*, 6656–6658.

[37] Dassonneville, B.; Witulski, B.; Detert, H. *Eur. J. Org. Chem.* **2011**, 2836–2844.

[38] Bagley, M. C.; Dale, J. W.; Merritt, E. A.; Xiong, X. *Chem. Rev.* **2005**, *105*, 685–714.

[39] Zou, Y.; Liu, Q.; Deiters, A. *Org. Lett.* **2011**, *13*, 4352–4355.

[40] Tanaka, K.; Suzuki, N.; Nishida, G. *Eur. J. Org. Chem.* **2006**, 3917–3922.

[41] Tanaka, K.; Hara, H.; Nishida, G.; Hirano, M. *Org. Lett.* **2007**, *9*, 1907–1910.

[42] Komine, Y.; Tanaka, K. *Org. Lett.* **2010**, *12*, 1312–1315.

[43] Kashima, K.; Ishii, M.; Tanaka, K. *Eur. J. Org. Chem.* **2015**, 1092–1099.

[44] Murayama, K.; Sawada, Y.; Noguchi, K.; Tanaka, K. *J. Org. Chem.* **2013**, *78*, 6202–6210.

[45] Xu, F.; Wang, C.; Wang, D.; Li, X.; Wan, B. *Chem. Eur. J.* **2013**, *19*, 2252–2255.

[46] Xu, F.; Wang, C.; Wang, H.; Li, X.; Wan, B. *Green Chem.* **2015**, *17*, 799–803.

[47] a. Wang, C.; Li, X.; Wu, F.; Wan, B. *Angew. Chem. Int. Ed.* **2011**, *50*, 7162–7166.
     b. Wang, C.; Wang, D.; Xu, F.; Pan, B.; Wan, B. *J. Org. Chem.* **2013**, *78*, 3065–3072.

[48] D'Souza, R.; Lane, T. K.; Louie, J. *Org. Lett.* **2011**, *13*, 2936–2939.

[49] Lane, T. K.; D'Souza, B. R.; Louie, J. *J. Org. Chem.* **2012**, *77*, 7555–7563.

[50] Richard, V.; Ipouck, M.; Merel, D. S.; Gaillard, S.; Whitby, R. J.; Witulski, B., et al. *Chem. Commun.* **2014**, *50*, 593–595.

[51] McCormick, M. M.; Duong, H. A.; Zuo, G.; Louie, J. *J. Am. Chem. Soc.* **2005**, *127*, 5030–5031.

[52] Stolley, R. M.; Maczka, M. T.; Louie, J. *Eur. J. Org. Chem.* **2011**, 3815–3824.

[53] Stolley, R. M.; Duong, H. A.; Thomas, D. R.; Louie, J. *J. Am. Chem. Soc.* **2012**, *134*, 15154–15162.

[54] Stolley, R. M.; Duong, H. A.; Louie, J. *Organometallics* **2013**, *32*, 4952–4960.

[55] Tekavec, T. N.; Zuo, G.; Simon, K.; Louie, J. *J. Org. Chem.* **2006**, *71*, 5834–5836.

[56] Kumar, P.; Prescher, S.; Louie, J. *Angew. Chem. Int. Ed.* **2011**, *50*, 10694–10698.

[57] Onodera, G.; Shimizu, Y.; Kimura, J.; Kobayashi, J.; Ebihara, Y.; Kondo, K., et al. *J. Am. Chem. Soc.* **2012**, *134*, 10515–10531.

[58] Satoh, Y.; Obora, Y. *J. Org. Chem.* **2013**, *78*, 7771–7776.

[59] Barluenga, J.; Jiménez-Aquino, A.; Fernández, M. A.; Aznar, F.; Valdés, C. *Tetrahedron* **2008**, *64*, 778–786.

[60] He, Y.; Guo, S.; Zhang, X.; Fan, X. *J. Org. Chem.* **2014**, *79*, 10611–10618.

[61] Bai, Y.; Tang, L.; Huang, H.; Deng, G.-J. *Org. Biomol. Chem.* **2015**, *13*, 4404–4407.

[62] Pan, X.; Liu, Q.; Chang, L.; Yuan, G. *RSC Adv.* **2015**, *5*, 51183–51187.

[63] Li, Z.; Huang, X.; Chen, F.; Zhang, C.; Wang, X.; Jiao, N. *Org. Lett.* **2015**, *17*, 584–587.

[64] a. Chun, Y. S.; Lee, J. H.; Kim, J. H.; Ko, Y. O.; Lee, S. *Org. Lett.* **2011**, *13*, 6390–6393.
b. Xun, Z.; Rathwell, K.; Lee, S. *Asian J. Org. Chem.* **2014**, *3*, 1108–1112.

[65] Wei, Y.; Yoshikai, N. *J. Am. Chem. Soc.* **2013**, *135*, 3756–3759.

[66] Abbiati, G.; Arcadi, A.; Bianchi, G.; Di Giuseppe, S.; Marinelli, F.; Rossi, E. *J. Org. Chem.* **2003**, *68*, 6959–6966.

[67] Bagley, M. C.; Dale, J. W.; Bower, J. *Chem. Commun.* **2002**, 1682–1683.

[68] Sakai, N.; Aoki, D.; Hamajima, T.; Konakahara, T. *Tetrahedron Lett.* **2006**, *47*, 1261–1265.

[69] Rieckhoff, S.; Hellmuth, T.; Peters, R. *J. Org. Chem.* **2015**, *80*, 6822–6830.

[70] Wang, Y.-F.; Chiba, S. *J. Am. Chem. Soc.* **2009**, *131*, 12570–12572.

[71] Wang, Y.-F.; Toh, K. K.; Ng, E. P. J.; Chiba, S. *J. Am. Chem. Soc.* **2011**, *133*, 6411–6421.

[72] Loy, N. S. Y.; Singh, A.; Xu, X.; Park, C.-M. *Angew. Chem. Int. Ed.* **2013**, *52*, 2212–2216.

[73] He, Y.-T.; Wang, Q.; Zhao, J.; Liu, X.-Y.; Xu, P.-F.; Liang, Y.-M. *Chem. Commun.* **2015**, *51*, 13209–13212.

[74] Manning, J. R.; Davies, H. M. L. *J. Am. Chem. Soc.* **2008**, *130*, 8602–8603.

[75] Safaei-Ghomi, J.; Ghasemzadeh, M. A. *J. Sulfur Chem.* **2013**, *34*, 23–241.

[76] Sridhar, M.; Ramanaiah, B. C.; Narsaiah, C.; Mahesh, B.; Kumaraswamy, M.; Mallu, K. K. R., et al. *Tetrahedron Lett.* **2009**, *50*, 3897–3900.

[77] Kantam, M.; Mahendar, K.; Bhargava, S. *J. Chem. Sci.* **2010**, *122*, 63–69.

[78] Tang, J.; Wang, L.; Yao, Y.; Zhang, L.; Wang, W. *Tetrahedron Lett.* **2011**, *52*, 509–511.

[79] Pagadala, R.; Kommidi, D. R.; Rana, S.; Maddila, S.; Moodley, B.; Koorbanally, N. A., et al. *RSC Adv.* **2015**, *5*, 5627–5632.

[80] He, X.; Shang, Y.; Yu, Z.; Fang, M.; Zhou, Y.; Han, G., et al. *J. Org. Chem.* **2014**, *79*, 8882–8888.

[81] Reddy, N. K.; Chandrasekhar, K. B.; Ganesh, Y. S. S.; Kumar, B. S.; Adepu, R.; Pal, M. *Tetrahedron Lett.* **2015**, *56*, 4586–4589.

[82] Banerjee, C. K.; Umarye, J. D.; Kanjilal, P. R. *Synth. Commun.* **2013**, *43*, 2208–2216.

[83] Zhu, C.; Bi, B.; Ding, Y.; Zhang, T.; Chen, Q.-Y. *Org. Biomol. Chem.* **2015**, *13*, 6278–6285.

[84] Raja, V. P. A.; Tenti, G.; Perumal, S.; Menéndez, J. C. *Chem. Commun.* **2014**, *50*, 12270–12272.

[85] Kantevari, S.; Chary, M. V.; Vuppalapati, S. V. N. *Tetrahedron* **2007**, *63*, 13024–13031.

[86] Kantevari, S.; Chary, M. V.; Vuppalapati, S. V. N.; Lingaiah, N. *J. Heterocycl. Chem.* **2008**, *45*, 1099–1102.

[87] Kumar, A.; Koul, S.; Razdan, T. K.; Kapoor, K. K. *Tetrahedron Lett.* **2006**, *47*, 837–842.

[88] Jiang, H.; Yang, J.; Tang, X.; Li, J.; Wu, W. *J. Org. Chem.* **2015**, *80*, 8763–8771.

[89] Dissanayake, A.; Staples, R. J.; Odom, A. L. *Adv. Synth. Catal.* **2014**, *356*, 1811–1822.

[90] Sim, Y.-K.; Lee, H.; Park, J.-W.; Kim, D.-S.; Jun, C.-H. *Chem. Commun.* **2012**, *48*, 11787–11789.

[91] Zheng, L.; Ju, J.; Bin, Y.; Hua, R. *J. Org. Chem.* **2012**, *77*, 5794–5800.

[92] Zhang, J.; Qian, H.; Liu, Z.; Xiong, C.; Zhang, Y. *Eur. J. Org. Chem.* **2014**, 8110–8118.

[93] Wan, J.-P.; Zhou, Y.; Cao, S. *J. Org. Chem.* **2014**, *79*, 9872–9877.

[94] Ren, Z.-H.; Zhang, Z.-Y.; Yang, B.-Q.; Wang, Y.-Y.; Guan, Z.-H. *Org. Lett.* **2011**, *13*, 5394–5397.

[95] Zhao, M.-N.; Hui, R.-R.; Ren, Z.-H.; Wang, Y.-Y.; Guan, Z.-H. *Org. Lett.* **2014**, *16*, 3082–3085.

[96] Yehia, N. A. M.; Polborn, K.; Müller, T. J. J. *Tetrahedron Lett.* **2002**, *43*, 6907–6910.

[97] Dediu, O. G.; Yehia, N. A. M.; Oeser, T.; Polborn, K.; Müller, T. J. J. *Eur. J. Org. Chem.* **2005**, 1834–1848.

[98] Kobayashi, T.; Hatano, S.; Tsuchikawa, H.; Katsumura, S. *Tetrahedron Lett.* **2008**, *49*, 4349–4351.

[99] Parthasarathy, K.; Jeganmohan, M.; Cheng, C.-H. *Org. Lett.* **2008**, *10*, 325–328.

[100] Parthasarathy, K.; Cheng, C.-H. *Synthesis* **2009**, 1400–1402.

[101] Parthasarathy, K.; Cheng, C.-H. *J. Org. Chem.* **2009**, *74*, 9359–9364.

[102] Hyster, T. K.; Rovis, T. *Chem. Commun.* **2011**, *47*, 11846–11848.

[103] Martin, R. M.; Bergman, R. G.; Ellman, J. A. *J. Org. Chem.* **2012**, *77*, 2501–2507.

[104] Chen, S.; Bergman, R. G.; Ellman, J. A. *Org. Lett.* **2015**, *17*, 2567–2569.

[105] Chen, X.; Wu, Y.; Xu, J.; Yao, H.; Lin, A.; Huang, Y. *Org. Biomol. Chem.* **2015**, *13*, 9186–9189.

[106] Too, P. C.; Noji, T.; Lim, Y. J.; Li, X.; Chiba, S. *Synlett* **2011**, 2789–2794.

[107] Colby, A.; Bergman, R. G.; Ellman, J. A. *J. Am. Chem. Soc.* **2008**, *130*, 3645–3651.

[108] Neely, J. M.; Rovis, T. *J. Am. Chem. Soc.* **2013**, *135*, 66–69.

[109] Kim, D.-S.; Park, J.-W.; Jun, C.-H. *Chem. Commun.* **2012**, *48*, 11334–11336.

[110] Ruiz, S.; Villuendas, P.; OrtuÇo, M. A.; Lledús, A.; Urriolabeitia, E. P. *Chem. Eur. J.* **2015**, *21*, 8626–8636.

[111] Zhang, Q.-R.; Huang, J.-R.; Zhang, W.; Dong, L. *Org. Lett.* **2014**, *16*, 1684–1687.

[112] Yamamoto, S.-i; Okamoto, K.; Murakoso, M.; Kuninobu, Y.; Takai, K. *Org. Lett.* **2012**, *14*, 3182–3185.

[113] Wu, J.; Xu, W.; Yu, Z.-X.; Wang, J. *J. Am. Chem. Soc.* **2015**, *137*, 9489–9496.

[114] Barluenga, J.; Fernández-Rodríguez, M. Á.; García-García, P.; Aguilar, E. *J. Am. Chem. Soc.* **2008**, *130*, 2764–2765.

[115] Ohashi, M.; Takeda, I.; Ikawa, M.; Ogoshi, S. *J. Am. Chem. Soc.* **2011**, *133*, 18018–18021.

[116] Chibiryaev, M.; De Kimpe, N.; Tkachev, A. V. *Tetrahedron Lett.* **2000**, *41*, 8011–8013.

[117] Hardegger, L. A.; Habegger, J.; Donohoe, T. J. *Org. Lett.* **2015**, *17*, 3222–3225.

[118] Liu, S.; Liebeskind, L. S. *J. Am. Chem. Soc.* **2008**, *130*, 6918–6919.

[119] a. Roesch, K. R.; Larock, R. C. *J. Org. Chem.* **2002**, *67*, 86–94.
b. Roesch, K. R.; Zhang, H.; Larock, R. C. *J. Org. Chem.* **2001**, *66*, 8042–8051.
c. Roesch, K. R.; Larock, R. C. *J. Org. Chem.* **1998**, *63*, 5306–5307.

[120] Diederen, J. J. H.; Sinkeldam, R. W.; Frühauf, H.-W.; Hiemstra, H.; Vrieze, K. *Tetrahedron Lett.* **1999**, *40*, 4255–4258.

[121] Chen, M. Z.; Micalizio, G. C. *J. Am. Chem. Soc.* **2012**, *134*, 1352–1356.

[122] Lei, C.-H.; Wang, D.-X.; Zhao, L.; Zhu, J.; Wang, M.-X. *J. Am. Chem. Soc.* **2013**, *135*, 4708–4711.

[123] Trost, B. M.; Gutierrez, A. C. *Org. Lett.* **2007**, *9*, 1473–1476.

[124] He, Z.; Dobrovolsky, D.; Trinchera, P.; Yudin, A. K. *Org. Lett.* **2013**, *15*, 334–337.

CHAPTER 4

Summary and Outlook

In conclusion, the main contributions on transitional metal-catalyzed synthesis of pyridines have been summarized and discussed. Most of the reported procedures are still based on complicated substrates, pre-preparation of starting materials is required. In the future, the development of methodologies based on using of simplified and readily available substrates is still under request.

Transition Metal-Catalyzed Pyridine Synthesis. DOI: http://dx.doi.org/10.1016/B978-0-12-809379-5.00004-4

Printed in the United States
By Bookmasters